GROUNDWORK GUIDES

Empire
James Laxer
Being Muslim
Haroon Siddiqui
Genocide
Jane Springer
Climate Change
Shelley Tanaka

Series Editor
Jane Springer

GROUNDWORK GUIDES

Climate Change

Shelley Tanaka

Groundwood Books
House of Anansi Press

Toronto Berkeley

Groundwood Books / House of Anansi Press
110 Spadina Avenue, Suite 801, Toronto, Ontario M5V 2K4
Distributed in the USA by Publishers Group West
1700 Fourth Street, Berkeley, CA 94710

ONTARIO ARTS COUNCIL
CONSEIL DES ARTS DE L'ONTARIO

We acknowledge for their financial support of our publishing program the Canada Council for the Arts, the Government of Canada through the Book Publishing Industry Development Program (BPIDP) and the Ontario Arts Council. Special thanks to the Ontario Media Development Corporation.

Library and Archives Canada Cataloging in Publication
Tanaka, Shelley
Climate change / by Shelley Tanaka.
(Groundwork Guides)
Includes bibliographical references and index.
ISBN-13: 978-0-88899-783-8 (bound)
ISBN-10: 0-88899-783-3 (bound)
ISBN-13: 978-0-88899-784-5 (pbk.)
ISBN-10: 0-88899-784-1 (pbk.)
1. Climatic changes–Environmental aspects. 2. Global warming. I. Title. II. Series.
QC981.8.C5T35 2006   363.738'74   C2005-907537-6

Design by Michael Solomon

Printed and bound in Canada

It is never too late to give up our prejudices...
What old people say you cannot do, you try
and find that you can. Old deeds for old
people, and new deeds for new.
— Henry David Thoreau

Contents

Chapter 1
Climate Change Is Here, and It's Real

There is an ecological time-bomb ticking away...
— Stephen Byers[1]

You've probably noticed that the weather has been wonky lately. In 2005, for example, Spain and Portugal suffered their worst drought in sixty years, threatening 80 million fruit trees, cutting the wheat harvest in half, and triggering deadly forest fires. Eastern North America baked under record-breaking heat waves, and there were blackouts in some places as people jacked up their air-conditioners. Europe had one of its hottest summers on record, an unsettling reminder of the tragic summer of 2003, when more than 30,000 people died from the heat.[2] Central America and central Canada were unusually wet, with heavy rains and floods. In the southern US, Hurricane Katrina caused devastating floods and record damage in New Orleans, a city that lies mostly below sea level. Some climatologists argued that a wetter, more intense tropical storm season could be due to global warming.

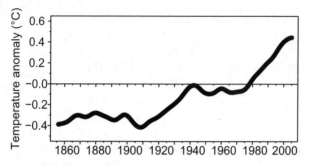

Global Temperature Record[3]

Average global temperatures have risen about 1°C (1.8°F) in the past 140 years (the horizontal line represents the average temperature from 1961 to 1990). In its 2001 report, the Intergovernmental Panel on Climate Change (IPCC) says most of the warming over the past fifty years is likely caused by an increase in greenhouse gases.

But perhaps the most eerie news is coming from the isolated wilderness of western Siberia, where researchers report that a gigantic expanse of permafrost — an area the size of France and Germany combined — has begun to melt for the first time since it was formed at the end of the last ice age. Icy ground that has been frozen for 11,000 years is now, in just a few years, turning into a landscape of mud and lakes, threatening to release huge amounts of methane, a powerful greenhouse gas.

Soggy northlands, heat-related deaths, water shortages, wilting crops, heavy precipitation, power blackouts.

Is this the wave of the future?

Many experts are saying Yes.

• • •

Scientists have been warning us about global warming for almost three decades. But many of us are only starting to get the message.

The planet is getting warmer, and the warming is largely being caused by human activity. More important, this warming is happening at an alarmingly rapid rate. The earth's surface temperature, which has not changed much in 10,000 years, has become significantly warmer during the past 150 years. If the current trend continues, many species, including humans, will not be able to adapt quickly enough to avoid severe hardship.

In the past hundred years, average global temperatures have risen at least 0.8°C (1.4°F), and three-quarters of this increase has occurred in the past thirty years. It doesn't sound like much, but consider this. At the depth of the last ice age 20,000 years ago — a time when ice covered most of Europe, and the island of Manhattan lay under a blanket of ice half a mile thick — the average global temperature was only 5°C (9°F) colder than it is now. And in the 100,000 years that humans have been around, the planet has

Warmest Years on Record[4]
1 2005
2 1998
3 2002
4 2003
5 2004

never been more than a degree or two warmer than it is today.

Besides, 0.8°C is just the *average* warming. The northern hemisphere is warming faster than the southern hemisphere. More dramatic change is taking place at the poles and in mountainous areas.

In 2002, on the eastern side of the Antarctic peninsula, a giant, floating mass of ice larger than the country of Luxembourg shattered and separated from the continent, disintegrating in just thirty-five days.[5]

At the other end of the globe, an area of Arctic sea ice one and a half times the size of Wales is lost each year, and an area of permanent sea ice the size of Arizona and Texas combined has disappeared since 1979. The Greenland ice sheet, up to 3 kilometers (2 miles) thick and just a little smaller than Mexico, is suddenly melting and sliding into the ocean much faster than scientists thought it would.

In fact, the World Glacier Monitoring Service says most of the world's glaciers are retreating. If the warming trend continues as expected, by 2050, Iceland will be virtually glacier free for the first time in at least 2 million years, polar bears could be extinct, and the Himalayan glaciers, which provide 500 million people with water, will be gone.

The world's fresh water is also at risk elsewhere, as lakes and rivers dry up, and as evaporation and seepage from rising sea levels leave higher concentrations of salt and pollutants in existing supplies. In the American Southwest, Lake Powell and Lake Mead have been drying up, threatening the water supply needed to run the Glen Canyon and Hoover dams.[6]

Widespread drying is occurring across much of Europe, Asia, Canada, Africa and Australia, doubling the area of the planet affected by drought in the past thirty years. Hot, dry conditions in China are causing intense sand storms that are blowing dust right across the Pacific, polluting the air on the west coast of North America.

Meanwhile, the upper parts of all the world's oceans are warming and expanding due to the rise in human-produced greenhouse gases.[7] Average sea levels have risen 10 to 25 cm (4 to 10 inches) in the past century, and they are expected to rise much more rapidly in the next century.[8] Warming sea temperatures are also contributing to the death of the world's coral reefs, the most biologically rich of all marine ecosystems.

Animals are changing their migration patterns. Butterflies and birds are flying north and breeding earlier in spring. Plant ranges are shifting toward the poles and up mountains. In the Canadian Arctic, people are spotting robins — a bird so rarely seen above the tree line that the local Inuit didn't even have a name for it. In the mountains of Colombia, warm-weather, disease-carrying mosquitoes that were once found only below 1,000 meters (3,300 feet) above sea level are now reported at altitudes more than twice that high.[9]

Each decade since the 1960s has been warmer than the last — an accelerated and extended warming that has not been seen in at least 1,200 years.[10] Climate changes that used to unfold over thousands of years are now happening over just a few decades.

And humans are to blame.

• • •

How can humans change the climate of the planet? There are many ways, but one of the most important is that in the past 150 years, as the world has become industrialized, people have been burning increasing amounts of carbon-containing fossil fuels — coal, oil and natural gas — releasing carbon dioxide into the atmosphere. The carbon dioxide traps heat and causes atmospheric temperatures to rise.

At the same time, we have been cutting down the earth's forests — forests that absorb carbon dioxide from the atmosphere and store it in their leaves and bark.

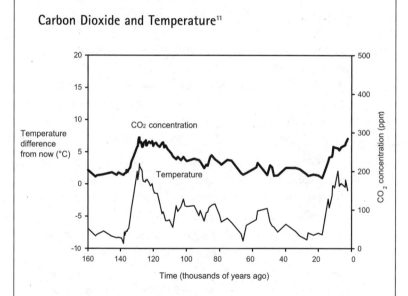

Carbon Dioxide and Temperature[11]

Ice core samples drawn from Antarctica show that temperatures and carbon dioxide levels have been in sync for at least the past 160,000 years. What will happen to global temperatures now that carbon dioxide levels are rising dramatically?

How Do We Know?

Today there are nine thousand weather stations, ships and buoys taking air and sea temperatures around the world. Since 1979, scientists have also been collecting temperature data from satellites circling the globe. But global temperatures have only been reliably recorded for about 150 years, a tiny period when it comes to tracking climate change.

So how do we know that temperatures and atmospheric carbon dioxide levels are rising? And how do we know that these two things are related?

The world's climate can be tracked over long periods in physical (proxy) evidence such as ice core samples. For example, researchers have drilled more than 3 kilometers (2 miles) into East Antarctica, pulling up long noodles of ice containing little bubbles of air trapped in snow that fell hundreds of thousands of years ago. Scientists can open up the bubbles and analyze the gases and chemicals inside to measure ancient carbon dioxide levels and estimate the temperature.

Ice core samples confirm that over hundreds of thousands of years, due to natural phenomena such as changes in the earth's orbit, atmospheric carbon dioxide levels have gone up and down, from 180 to 280 ppm (parts per million). And as carbon dioxide levels rose and fell, so did the average global temperatures. Ice cores also show that after the mid 1700s (the beginning of the use of the steam engine), carbon dioxide levels began to rise beyond 280 ppm, and they have been steadily rising since, up to about 380 ppm today. They are now 27 percent higher than at any point in the past 650,000 years,[12] and they are rising at an unusually fast rate.

Climate change can also be seen in other proxy evidence. Silt and stone taken from deep sea bottoms can show, for example, where icebergs once drifted and melted, depositing their sediments on the ocean floor. Fossilized beetles (insects that are especially sensitive to climate changes) can be identified as hot- or cold-loving species to determine what kind of climate they once thrived in. Pollen grains taken from ancient swamps can show that heat-loving trees once grew there, and when. Carbon dioxide levels can also be measured by counting the number of leaf pores in ancient tree leaves preserved in peat bogs or ice, or by measuring tree rings and the bands on corals.[13]

With Our Own Eyes

Although global temperatures have only been systematically measured since 1861, humans have been recording evidence of climate change for thousands of years. In the Sahara, ancient rock carvings depicting a fertile landscape full of browsing animals like giraffes are potent evidence that what is now desert was once covered with lush vegetation. Six-thousand-year-old drawings of hippos — animals that need water year round — have been found in areas that are now arid. Early written accounts from ancient Sumeria, China and Egypt tell us whether there was enough rain or sun to grow certain crops, whether the game animals were healthy, and when extreme cold, heat or storms wiped out local communities.

Why They Are Called Fossil Fuels

Coal, oil and natural gas are called hydrocarbons because they are composed mainly of hydrogen and carbon. These fuels were formed as long as 300 million years ago. Ancient trees and plants grew and died in swamps where there was not enough oxygen for them to decay. Instead, they, and the carbon they contained, were buried and compressed, eventually forming coal. Oil and natural gas were formed from the bodies of millions of very small dead sea plants and animals that sank and accumulated at the bottom of the ocean, which is why some oil deposits are found underwater.

More than half of the world's forests have been cut down in the past 8,000 years, and each year an area larger than the state of Florida is destroyed.[14]

Each day we burn such vast quantities of fossil fuels that the carbon that has been stored underground for hundreds of millions of years is being returned to the atmosphere within a few centuries. There is now 30 percent more carbon dioxide in the atmosphere than there was 250 years ago. And more is released every time humans burn fossil fuels to heat buildings, make electricity or drive their cars.

Who is producing all this carbon dioxide? The US and China are the biggest overall emitters by country. However, per capita, the biggest emitters are Americans, Australians and Canadians (the average American, for

Carbon Dioxide Emissions Per Capita[15]	
US	20.0 metric tons
Australia	18.4 metric tons
Canada	18.4 metric tons
Japan	9.8 metric tons
Great Britain	9.1 metric tons
France	6.8 metric tons
Sweden	6.1 metric tons

If each person on the planet were allowed roughly the same amount of carbon dioxide emissions, that amount would be slightly more than one metric ton of carbon. At present, the average person in developing countries produces half this much. The average American and Canadian produces as much carbon dioxide as 18 Indians or 99 Bangladeshis.[16]

example, consumes six times as much energy as the average person on the planet[17]).

But the atmosphere has no borders, and what goes into the atmosphere in one part of the world quickly affects us all.

• • •

Warming Around the World[18]

More drought and forest fires in western Canada than in 400 years

Summer ice breakup on western Hudson Bay occurs 2 weeks earlier than 20 years ago

In Britain frogs mate 7 weeks earlier than in 1950

More than 80% of Atlantic beaches eroding

In the US, more than 45 million live on hurricane-prone coastlines

Heat-loving fungus threatens Costa Rican frogs with extinction

Glaciers in Argentina disappearing at a rate of 10,000 football fields per year

Average winter temperatures over Antarctica have risen 2°C (9°F) in 30 years

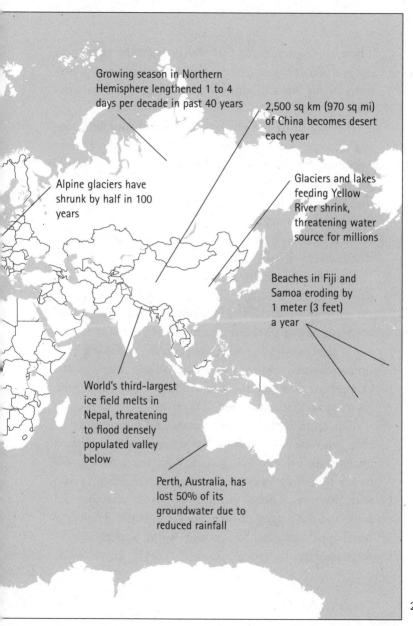

Growing season in Northern Hemisphere lengthened 1 to 4 days per decade in past 40 years

2,500 sq km (970 sq mi) of China becomes desert each year

Alpine glaciers have shrunk by half in 100 years

Glaciers and lakes feeding Yellow River shrink, threatening water source for millions

Beaches in Fiji and Samoa eroding by 1 meter (3 feet) a year

World's third-largest ice field melts in Nepal, threatening to flood densely populated valley below

Perth, Australia, has lost 50% of its groundwater due to reduced rainfall

There are many natural forces that cause the climate to get warmer, ranging from volcanic activity to fluctuations in the sun's brightness and changes in the earth's orbit.

But one thing is now clear. Natural phenomena cannot account for the speed and extent of the current warming. Scientists estimate that these natural forces have contributed only one-quarter of the total warming that took place in the twentieth century.[19] The rest must be blamed on human activity.

So if humans have caused the problem, it makes sense that humans should fix it.

But who is going to do it? Global warming has been called the greatest single challenge facing humanity, yet the world's political and business leaders seem to be spending much of the time finding reasons not to act. Some say that dealing with the problem will cause economic disaster. Others claim that we can adapt to the changes, or new technologies will save us, or at least the smartest and richest of us will survive.

The truth is that the politicians and business executives who are now making the decisons about climate change will be dead and gone in fifty years. And it will be left to today's young people and their children to deal with the mess. Global warming is already affecting our lives, and the biggest changes are yet to come. They may not happen in time for current world leaders to take responsibility for them. Instead, they'll arrive when today's young people are looking for jobs, houses, raising families.

The climate change issue is tough to face because it goes far beyond agreeing on the science. The scientific consensus is clear. But global warming is an economic, political, philosophical and emotional issue as well. It affects how governments form policy and their relationship to industry.

And it affects the choices we make in our daily lives, how we look at the rest of the world, and what kind of world we want to leave to future generations.

Chapter 2
How We Got Here

Progress, man's distinctive mark alone...
— Robert Browning

Over the earth's long lifetime, the climate has changed many times, for many reasons. Continents have broken up and shifted, affecting the movement of warm and cool ocean and air currents. Periods of heavy volcanic activity have filled the atmosphere with dust and gases, blocking out the sun. Changes in the sun's brightness and the earth's orbit and tilt have influenced the amount of sunlight reaching the earth.

All these forces have caused natural long-term cycles of warming and cooling that have taken place over hundreds, thousands or millions of years.

About 90 million years ago, for example, the ocean was warmer than a hot tub.[1] During a particularly warm period 55 million years ago, there were subtropical forests in the Arctic, and the ancestors of today's crocodiles dozed in swamps in northern Europe.

Over the past 2.5 million years, the earth has also experienced several long periods of cooling. During these cold periods, or ice ages, a good deal of the earth's water turned into snow and ice. Sea levels fell more than 100 meters (300 feet) as the polar ice sheets thickened and spread over the surrounding land masses, covering much of North America and Europe with glaciers. These ice ages (the most recent began 120,000 years ago) have been interrupted by shorter periods of warming (inter-glacials), when the edges of the ice sheets melted, and ocean levels rose.

These cycles of warming and cooling have always affected the plants and animals on the planet. Dinosaurs, for example, were in decline due to climate change long before an asteroid finally wiped them out 65 million years ago. As different parts of the world warmed and cooled, some species moved or adapted (at one point when the world was warmer than it is today, elephants and hippos thrived in western Europe). Others, like the mammoth and saber-toothed tiger, died off, while new species, such as the polar bear, emerged.

Modern humans have only been around for 100,000 years or so but, like other species, humans adapted as the planet warmed and cooled. In the early days, people moved to where the food was. Small groups of nomadic hunters might follow herds of reindeer or bison as the climate changed and grazing territories shifted. People lived where nuts, berries, grasses, fungi and shellfish were plentiful. If their traditional food sources disappeared, they moved to find others. Groups that could not find enough to eat simply died off, and the human population remained small.

Then, about 10,000 years ago, humans began to farm. Instead of roaming the land in search of food, people began to stay in one place, where they grew food crops and raised animals such as cattle, goats, sheep and pigs. They found ways to store and preserve food —

How We Got Here: A Few Highlights

10,000 BC	People begin to farm in southeastern Asia
1750s	Industrial Revolution begins
1765	James Watt redesigns the steam engine to make it more efficient, ushering in widespread use
1800	3 percent of the world's population lives in cities
1850s	Widescale manufacture of steel begins
1859	First oil well drilled in Titusville, Pennsylvania
1900	14 percent of the world's population lives in cities
1908	Henry Ford manufactures his first automobile; by 1913 his company is manufacturing 1,000 cars a day on assembly lines
1960s	Oil replaces coal as world's main source of energy
2000	Half the people in the world live in cities

grinding grain into flour to make bread, salting fish and meat, boiling nuts and grains to make oil.

Groups of people settled in places where the climate was mild, because warmer, longer growing seasons meant bigger, more reliable crops. Settlements sprang up along low-lying shorelines on rivers and oceans, where there were natural harbors and easy access to water transportation and irrigation. Once people had a more reliable food supply, populations began to grow rapidly in different parts of the world.

As settlements grew and spread, surrounding forests were cut down for fuel for cooking and heating, to clear land for towns and roads, and for crops and grazing. In many places, wood was used to build homes, churches and schools (it took twelve mature trees, for example, to make a single house[2]), as well as carts, furniture, fortifications, bridges and, eventually, ships, railway ties and paper.

When the trees and other resources around them ran out, people looked farther afield. Europeans went to the Americas, to Russia. By the twentieth century, widespread clearing was flattening the tropical forests of Southeast Asia, South America and Africa.

As the world's population grew, more and more land was needed for farming on a larger scale — sugar plantations in the Philippines, coffee farms in Brazil, sheep pastures in New Zealand, tobacco fields in the United States, logging operations in Indonesia and Thailand.

When several seasons of intensive farming had exhausted all the nutrients in the soil, farmers simply moved on and cleared more land.

Over time, the world's forests began to disappear. Three-quarters of China was originally covered by trees; now only 5 percent of the land is forested. In the United States, all but 6 percent of the original forest cover was cleared.[3] Brazil, home to much of the planet's remaining rainforest, has already lost half of its major old-growth forest and continues to lose 1.6 million hectares (4 million acres) of forest a year.[4] Half the world's mature tropical forests were cut after World War II, mostly to make room for agriculture.[5] Often the clearing was achieved by burning, which was faster than cutting. When trees were burned, the carbon stored in their leaves and trunks was released into the air.

As wood fuel became scarce, people turned to coal, a carbon-rich fossil fuel formed hundreds of millions of years ago from decomposing trees and swamp plants. Coal was plentiful, and it could be dug out of the ground and burned to heat homes and cook food.

In 1765, a British engineer named James Watt built an engine that ran on steam. The steam came from boiling water heated by burning coal. The engine could be used to run trains, ships and machinery. Coal could also be converted into gas to light lamps. Now factories could be lit at night, and people could work longer hours. In cities, smoke from burning coal spewed out of home and factory chimneys twenty-four hours a day. The smoke settled in the air in a yellow haze called smog. The burning coal also released large amounts of carbon dioxide into the atmosphere.

New machines were invented. Agriculture became mechanized with the use of mechanical reapers, combine harvesters, seed drills and plows. Farms became larger, more food could be grown and more people could be fed, but fewer people were needed to work in the fields. Instead, people moved to the cities to work in the growing numbers of factories. This period of large-scale industrialization was called the Industrial Revolution, and it changed how people lived — first in Europe, then in North America and later in other parts of the world.

As the Industrial Revolution progressed, more and more energy was needed — to turn iron into a harder metal called steel, to smelt copper, tin and lead, to make glass, bricks and eventually new materials like aluminum and plastics. Cement (made from heated limestone and then mixed with sand and broken stones to form concrete) became the most widely used building material in the world, but huge amounts of fossil fuels were burned to make it.[6]

Energy was needed to power ships that carried raw materials to the factories to be turned into clothing, processed foods and other goods — cotton from the American South, furs from Canada, beef from Argentina, rubber from the tropics, wool from Australia.

Meanwhile, new appliances and devices were invented to make life easier, but all these things consumed energy, too, both when they were produced and when they were used. More energy was needed to power refrigerators, stoves, washing machines and dryers; then to run televisions, air-conditioners, dishwashers, stereos and computers.

As the need for energy grew, oil, another fossil fuel, gradually replaced coal as the world's most important source of energy. Oil was used to produce electricity and then to fuel automobiles after the internal combustion engine was perfected, and cars replaced horses as the main mode of transportation. Cars didn't need to be stabled and fed and they didn't leave a mess on the roads, but they did run on gasoline — refined oil — and they released carbon dioxide whenever they were driven.

Cars became the center of North American culture, influencing how and where people lived. As cities grew, urban planners designed suburbs — satellite communities where every house had room to park an automobile or two, but people had to get in their cars to get to their workplace or to go shopping. Nobody minded, because gas was so cheap. Soon people came to expect to be able to drive their own cars to work, to stores, even to school, creating a need for more expressways, bridges, underpasses, overpasses, parking lots and wider streets.

In the beginning, most of this development occurred in Europe, North America and Japan, as these industrialized countries drew first on the raw resources and then on the markets of the rest of the world. These Western economies continued to grow richer, while most of the world remained poor. Even today, more than a billion people in the world do not have clean water or proper housing or enough fuel to heat their homes, while people in the industrialized world, who make up only one-quarter of the population, eat half the world's food.[7]

But some developing nations are catching up quickly. China, the world's most populous country, now has the fastest-growing economy. It is copying the patterns of Western consumption at an accelerated rate, and the model that took the industrialized world many decades to achieve — large homes in suburbs, expanding cities and highways and a car-centered culture — is now being created at a breakneck pace in some parts of China.

World Population Then and Now[8]

10,000 BC	10 million
AD 1	300 million
1750	760 million
1800	1 billion
1930	2 billion
1960	3 billion
1990	5 billion
2000	6 billion
2006	6.5 billion

Meanwhile, the population of the world has continued to grow. Now, millions of people are affected when droughts turn over-farmed land into dustbowls, when heavy rains turn clearcut hillsides into mud, when floods engulf heavily populated river deltas.

Modern human lifestyles now depend on consuming vast amounts of energy — to keep billions of people sheltered and warm; to produce, refrigerate, preserve and transport food; to carry humans and cargo from place to place; and to produce the many goods that fill people's daily lives. The world's energy use has nearly doubled in the past thirty years, and is expected to increase 60 percent by 2020.[9]

And although hydro power, nuclear energy and other sources carry some of the burden, the vast majority of the world's energy still comes from burning fossil fuels.

Chapter 3
How the Climate System Works

> Each element of the cosmos is positively woven
> from all the others...It is impossible to cut into
> this network, to isolate a portion without it
> becoming frayed and unravelled at all its edges.
> — Pierre Teilhard de Chardin

The earth's climate is created by the complex relationships that exist between the sun, atmosphere, water and land. Their interactions determine everything from temperature and precipitation to wind, humidity and atmospheric pressure — whether it is hot or cold, windy or calm, wet or dry.

Sun

Everything begins with the sun. When sunlight hits the earth, some of it is reflected back into space. But some is absorbed by the earth's surface and turned into heat — heat that sets both the atmosphere and oceans in motion, triggering all of the earth's climate systems.

Atmosphere

Although the atmosphere has often been described as a blanket of air surrounding the earth, it is an extremely thin blanket — as thin, say, as the skin on an apple. It contains mostly nitrogen (78 percent) and oxygen (21 percent). Many other gases make up the remaining 1 percent. Some of these gases are only present in tiny amounts — parts per million (ppm) or even parts per trillion — but together they create a natural phenomenon called the Greenhouse Effect.

What Are Greenhouse Gases?

Dozens of gases are responsible for the Greenhouse Effect. Water vapor makes up the largest part, and on average a water molecule only stays in the atmosphere for a week or so. However, its total amount remains stable, since it is recycled over and over in the hydrologic cycle.

Carbon dioxide is the most significant greenhouse gas directly released into the atmosphere by humans. Atmospheric levels are steadily increasing, because each molecule can remain in the atmosphere for a hundred years or more. Carbon dioxide is responsible for at least half of human-created global warming.

Methane, another greenhouse gas, is the main chemical in natural gas. It is also produced when organic matter decays in the absence of oxygen, in places such as thawing tundra, swamps, landfills containing rotting garbage, and rice paddies. (Rice is the staple food for half the people in the world, and rice fields may be the biggest human-created source of

In a greenhouse, sunlight passes through the glass and is absorbed by the plants and soil, which then give off heat. The heated air would normally rise and be replaced by the cold air from above, but the glass prevents this, keeping the air inside the greenhouse warm.

Similarly, sunlight passes through the earth's transparent atmosphere, where it is absorbed by the earth's surface, converted into heat and emitted back into the atmosphere. Greenhouse gases trap some of this heat, and this heat warms the earth.

methane, producing up to 25 percent of global methane emissions.[1] Methane amounts have doubled since 1800, largely because of the increase in rice paddy agriculture as the world's population grows.[2]) Methane is also produced by farm animals as they digest food. It is far less plentiful than carbon dioxide, but it is at least twenty times more effective at absorbing heat and reradiating it back to earth, and each molecule hangs in the atmosphere for about twelve years. It is responsible for about one-quarter of human-caused warming.

Other greenhouse gases include nitrous oxide, chlorofluorocarbons (CFCs) and ozone. All these gases exist in the atmosphere in very tiny amounts, but we need to keep an eye on them. Nitrous oxide, for example, which comes out of vehicle tailpipes and is produced in some fertilizers, is only responsible for 6 percent of human-caused warming, but it is 200 times more powerful than carbon dioxide as a greenhouse gas, and it can stay in the atmosphere for more than one hundred years.

If there were no greenhouse gases trapping the outgoing heat, the earth's surface temperature would be -18°C (0.4°F) — too cold for life as we know it.

When the atmosphere at the equator is warmed by the sun, the hot air rises and moves part way toward the poles. As the air rises and moves, it cools, becomes denser and sinks, eventually moving back toward the equator.

This movement of cold and hot air does not happen smoothly or evenly. The size and shape of land masses, the rotation of the earth on its axis, and countless other influences all affect the speed and density of the moving air masses and how they meet each other. When two air masses meet, changes in the weather occur.

Water

The sun's energy also heats the oceans, which cover 70 percent of the earth's surface. The heated water sets ocean currents in motion the same way heated air creates the winds. Currents carry warm ocean surface water from the tropics toward the poles, while denser, colder deep water moves in to take its place. Contrasting temperature and salinity (which together determine density) keep the water masses separate.

The ocean currents move like a giant conveyor belt that circles the globe, carrying warm water from the equator up the east coast of North America and across the Atlantic as the Gulf Stream. When this warm water reaches the mid North Atlantic, it releases its heat into the atmosphere, cools, becomes denser and sinks. The deep cold water eventually flows back south, around the southern tip of Africa and into the Pacific, where it warms again and finally makes the return trip west and back to the Atlantic.

Thermohaline Circulation[3]

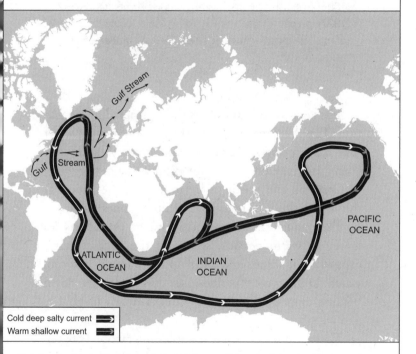

Cold deep salty current ➡
Warm shallow current ➡

It takes a drop of water a thousand years to complete the journey along the giant conveyor belt known as the thermohaline circulation.

The giant loop is known as the thermohaline circulation (THC) — a massive, slow, steady movement of water that has a powerful influence on the world's climate. Thanks to the Gulf Stream, for example, palm trees grow in Cornwall and southern Ireland, which lie almost at the same latitude as the southern tip of James Bay.

The temperature of the ocean waters also affects sea levels. In warm periods, sea levels rise because water expands as it warms and takes up more room than cool water. During cold periods, sea levels also go down because more of the planet's water is frozen in glaciers and ice caps.

There is also a close and complex interaction between the earth's bodies of water and the atmosphere. This interaction is called the hydrologic cycle.

Most of the earth's water (97 percent) is held in the oceans. Three-quarters of the rest lies frozen in the polar ice caps and in mountain glaciers; most of the remainder is buried underground. Together, freshwater lakes and rivers and water vapor in the air make up only about one hundredth of one percent (.01 percent) of all the water on earth.

When the sun heats the oceans, lakes and rivers, the water evaporates and rises into the air as water vapor. As

Smog
A Separate Issue... Sort of

The term smog was first used to describe the air pollution created by coal-burning factories in London in the early 1900s (smog = smoke + fog). Today it refers to the smoke that comes out of factory chimneys and vehicle tailpipes, forming a stagnant yellow cloud that sits over large cities on days when there is no wind. The chemical-laden air is hard to breathe and can trigger health problems such as asthma.

Smog (and other kinds of air pollution such as acid rain) is not directly connected to carbon dioxide emissions. However, it is related to global warming, because it is also created by burning fossil fuels, and because reducing greenhouse gas emissions will also improve air quality by reducing smog.

Smog can be controlled by removing certain chemicals from gasoline, and by adding catalytic converters to cars to remove some of the chemicals. Catalytic converters do not, however, get rid of carbon dioxide, the most problematic greenhouse gas.

the air rises and cools, the water vapor condenses, releasing heat as it does. As it condenses it also collects on small particles (aerosols) that are suspended in the atmosphere. If there are enough drops of water vapor, clouds form. If the water drops become large enough, clouds become saturated, and the water falls to the earth as rain or snow, replenishing the world's lakes, rivers and oceans, groundwater and ice caps.

At any given moment, clouds cover more than half the planet, and they affect the earth's climate in many different ways. Some low gray clouds produce precipitation. Wispy, high clouds let in light but reduce the amount of heat that can get back out. And low-lying clouds have a cooling effect, because they trap essentially no heat, and they bounce the sun's light energy back into space before it even reaches the earth.

The amount of water on earth does not really change. Instead, the same water is recycled over and over. But the hydrologic cycle does not work tidily, and the water seldom goes up and down in the same place. Winds can move clouds in unpredictable ways, and the water that evaporates off a lake can fall as rain a half a continent away.

Land

The earth's land areas also affect climate in different ways. Temperatures are more extreme, for instance, in the middle of large land masses. Mountains force the air

Ozone
A Separate Issue... Sort of

Ozone (O_3) is a form of oxygen. It is concentrated in the upper atmosphere (stratosphere), where it absorbs and thus blocks some of the sun's incoming ultraviolet rays. This is a good thing, because too much ultraviolet radiation can damage the DNA molecules in living things, causing skin cancer and eye cataracts in humans as well as damaging the soft tissues of creatures like frogs and seals. Ultraviolet radiation can kill the plankton that form the base of the ocean food chain. It can also damage food crops such as soybeans.

In the mid 1900s, artificial chemicals called chlorofluorocarbons (CFCs) were invented to puff up liquid plastic into plastic foam, as refrigerants in fridges and air-conditioners, and as propellants in aerosol spray cans. As CFCs escaped into the air, they floated up into the stratosphere, where they broke apart the ozone molecules, allowing more ultraviolet radiation to reach the earth.

When scientists discovered there was a large hole in the ozone layer above Antarctica, many countries agreed to cut their production of CFCs, and ozone levels are now starting to recover.

The thinning of the ozone layer and global warming are connected in a few ways, and the relationship is complicated. For one thing, while global warming causes the lower atmosphere (troposphere) to get warmer, it also causes the stratosphere above it to get cooler, and a cooler stratosphere speeds up ozone destruction in the polar regions.

Ozone and CFCs are also both greenhouse gases, though they are present in the lower atmosphere in very small amounts.

that hits them to rise so that the water vapor condenses into clouds, which release their water as rain and snow.

The color of the land affects climate, too. Just as a black car sitting in the sun will feel hotter than a white one, dark-colored bare rock will soak up the sun's rays and turn the energy into heat; such surfaces are said to have low albedo. Light-colored areas of ice and sand have high albedo. They reflect about 40 percent of sunlight back into space before it can warm the surface. Freshly fallen snow has the highest albedo of all, reflecting 80 percent or more of the light that hits it back into space before it can turn into heat.

Carbon Cycle

The sun's effect on the air, water and land creates the earth's climate. But how are humans disturbing this natural system?

Humans are affecting climate by changing the balance of another vital natural cycle called the carbon cycle.

Carbon is a chemical element found in everything from plants and animals (one-fifth of the human body, for example, is carbon) to certain rocks and the air we breathe. Most of the planet's carbon is buried in the earth's crust as fossil fuels and embedded in rocks like limestone and chalk. Carbon is also stored in the earth's oceans, soils and plants.

Carbon constantly circulates between the air, land,

water and all the living things on the planet. Carbon (C) is introduced into the atmosphere as carbon dioxide (CO_2). This happens naturally in a number of ways. Volcanoes bring up carbon from deep below the earth's surface when they erupt. Limestone rocks erode slowly over time, releasing carbon into the air.

And all living things contribute to the carbon cycle when they breathe in oxygen, combine it with the carbon in their bodies, and breathe it out again as carbon dioxide. When plants and animals die, their bodies decompose and release their carbon back into the soil, air or water.

Much of the carbon released into the air is naturally recycled through the world's vegetation and oceans, which pull carbon out of the atmosphere.

Plants take in carbon from the air and use it to make leaves and wood. All types of vegetation take in carbon, but young, growing forests, especially tropical forests (which grow year round), are the most efficient.

The oceans have a much larger capacity to take up carbon; they absorb one-third of the carbon dioxide that humans put into the atmosphere. Waves and other turbulence mix carbon dioxide from the atmosphere into the water. Some of the carbon dioxide is used by tiny marine plants (phytoplankton) for growth and by ocean animals to build their shells. When these plants and animals die, they sink into deep water where the carbon can be sequestered for hundreds or even thousands of years.

Carbon by Numbers[4]

Atmospheric carbon dioxide is measured as parts per million (ppm) by volume. If, for example, the measurement of carbon dioxide in the atmosphere is given as 379 ppm, it means that for a million liters of atmosphere, there are 379 liters of carbon dioxide.

Carbon dioxide emissions (e.g., the amount produced by cars, or by burning fossil fuels) are usually expressed as metric tons.

Carbon Dioxide in the Atmosphere

During ice ages	180-220 ppm
Between ice ages	260-280 ppm
1750	280 ppm
1958	315 ppm
1992	355 ppm
2006	380 ppm

Amount of carbon in the atmosphere: 50 billion metric tons

Amount of carbon held in the world's vegetation, including forests: 600 billion metric tons

Amount of carbon held in the world's oceans: 40,000 billion metric tons

Amount of carbon held in 1 square meter of rainforest (plants and soil): 25 kilograms

Amount of carbon held in 1 square meter of cow pasture: 4 kilograms

Amount of carbon held in 1 square meter of desert: less than 1 kilogram

Amount of carbon put into the atmosphere each year by humans burning fossil fuels: 7 billion metric tons

And if these ocean bottom accumulations are buried in mud, they can be removed from the earth's surface systems for millions of years.

Like all of the earth's natural systems and cycles, the carbon cycle has been operating for millions of years. But this system is now being influenced by a new agent. Humans are pulling huge amounts of carbon fuels — oil, coal and natural gas — out of the ground and burning them, releasing large amounts of carbon dioxide into the atmosphere over a very short period. At the same time, we have been rapidly cutting and burning the earth's forests, adding carbon dioxide to the air while reducing the forests' ability to remove excess carbon from the air.

Carbon dioxide is now accumulating in the atmosphere twice as fast as the natural sinks can remove it, causing the planet to get warmer. And the effects of this warming are long-term and far-reaching.

Chapter 4
The Effects of Climate Change

It is almost as if we had lit a fire to keep warm, and
failed to notice that the furniture had ignited.
— James Lovelock[1]

Temperature readings, ice core samples and other physical evidence all show that average global temperatures are rising. But rising temperatures, particularly of a degree or two, don't mean much on their own. This week is a little warmer than last week. This winter is milder than last winter. So what? Maybe it's a good thing.

But warmer global temperatures affect more than the daily or monthly weather forecasts. They have an impact on everything from the world's shorelines and the planet's fresh water supply to where we live, what we eat, the jobs we work at — even the survival of some species.

The interconnectedness of the oceans, land and atmosphere means that the effects of global warming will be complex, far-reaching and even contradictory. Some places will be a little warmer, some a lot warmer, and some may at first even be cooler, if winds or ocean currents shift. Wind and rainfall patterns that have remained more or less constant for thousands of years may now change within decades. Worldwide there will be more precipitation, but shifting wind patterns may carry clouds long distances before they dump their water vapor. To the poles, where their moisture will fall as snow, replenishing the polar ice sheets? To areas that were once dry, opening up new farmland? In a warmer atmosphere will this rainfall evaporate before it can benefit soils and plants? Will areas that once received rain now go dry?

No one knows for sure.

Effects will breed other effects, or feedbacks. Some of these feedbacks may reduce the existing warming (negative feedbacks). Warmer temperatures, for instance, might allow plants to grow faster, taking more carbon dioxide out of the atmosphere. Positive feedbacks increase or accelerate the warming. Once ice and snow begin to melt, for example, the melting ice absorbs more sunlight. If a warmer climate causes forests to dry out and die, they will no longer remove carbon from the atmosphere, which will cause the atmosphere to heat up even more.

Here are just some of the things that happen when global temperatures go up.

THE CLIMATE BECOMES MORE VARIABLE

A warmer climate may trigger more episodes of extreme weather, such as heat waves, storms and drought. Although no single weather event can be directly blamed on global warming (that event may have occurred as a result of natural forces), a warmer climate does make extreme weather events more likely to happen.

Heat Waves

When average temperatures go up, cold days are not as cold, but extremely hot days are even hotter, and there are more of them. People lucky enough to have air-conditioning will want to turn it on more often, but many will suffer, especially the poor, the elderly and the ill. Death rates typically double or triple in large cities during very high temperatures, because asphalt, concrete, brick and glass soak up and hold heat and because heat releases allergens such as pollen and mold, triggering conditions such as asthma, especially in children.[2] The 2003 heat wave in Europe, when temperatures went over 40°C (104°F) in some places, resulted in more than 30,000 heat-related deaths.

More Precipitation

A warmer planet is a wetter planet. It is estimated that global precipitation increases 1 percent for every 1 degree Celsius of warming. The total amount of water on the planet and in the atmosphere remains the same, but the hydrologic cycle becomes more active. More water evaporates, and water vapor rises and falls as rain more frequently.

The combination of a moister, warmer atmosphere and warm ocean temperatures provides the energy that fuels tropical storms, and some climatologists now think there may be a connection between tropical storms and global warming.

Global warming does not mean there will be more tropical storms.[3] But the ones that do occur may well be wetter and more intense, and this trend will likely continue as the climate continues to warm.[4] Natural long-term cycles drive the hurricane season in the mid Atlantic, El Niño events in the Pacific and the monsoon season in Southeast Asia. However, such storms get their energy from warmer surface ocean temperatures. The extra heat and energy trapped by greenhouse gases heat up the atmosphere even more. Winds rush in at ground level to replace the rising air, and wind speeds increase, violently whipping the rain and waves.

More rain also means more heavy rain.[5] When rain comes down in sudden, intense gushes, the water doesn't soak into the soil but runs off, often taking the nutrient-rich topsoil with it. Sewers, rivers and other bodies of water overflow, causing flooding. Once the storm has passed, mosquitoes are drawn to the pools of leftover warm stagnant water, with some species bringing diseases such as malaria, encephalitis and dengue fever.

Crops are drowned and flattened. Power lines and trees come down. Bridges and roads are swept away and flooded. The drinking water is polluted from overflowing sewage systems. Slabs of wet earth fall off the sides of mountains, causing landslides. Rivers flowing from high areas bring huge amounts of water and debris downstream, flooding river deltas. Rising sea levels mean high waves reach farther inland and, because of growing populations, do more damage to areas that were once safe from storms.

In the aftermath of such storms it can take years for people to rebuild their homes and recover their livelihoods.

Drought

It's true that a warmer climate means a more active hydrologic cycle and more precipitation. But excessive precipitation in one place means less somewhere else. As well, with higher temperatures, moisture evaporates more quickly from plants and the soil, so drought can occur even if the rainfall remains unchanged. Warmer, shorter winters can mean less snow accumulation, so a spring melt that normally replenishes lakes and rivers may not provide enough water for the next growing season. Global warming does not *cause* drought; some areas are naturally drought-prone. But it can make droughts longer and more intense.

Food plants, which are usually seeded and harvested

in one growing season, are particularly vulnerable to drought. But drought kills plants in more ways than by not providing the water they need to grow. On the North American prairies, for example, a warm, dry spring can mean emerging grasshopper nymphs are not killed by the usual spring rains or damp-loving fungal infections, resulting in an exceptionally large grasshopper population that can decimate wheat crops. Spider mites can invade soybeans if the insects are not killed by the damp molds that normally keep them in check.

Trees may have deep roots and a tough armor of bark, but drought can stress even hardy species, making them vulnerable to insects and disease. Bark beetles not killed off by freezing winters move in to feed on cedar, fir and pine that have been weakened by dry conditions. Flatheaded borers go after the oaks and birches.

And, of course, drought turns forests into tinder boxes, vulnerable to fires that affect all species that live in forests or depend on them for livelihood. The loss of large areas of forests to fires also affects the atmospheric carbon dioxide level, adding carbon to the air as the wood burns.

Periodic droughts are natural and normal, especially when they occur in the middle of large land masses. A drought involves one or two seasons of below-average precipitation. In farming country, that season's crops will be poor or fail, but the farms will recover when the rain replaces the moisture in the soil the following year.

If there is low rainfall for a number of years in a row, however, as happened during the prolonged drought of the 1930s on the North American prairies, the result is desiccation. Soils dry out, even deep-rooted trees may die, and the vegetation may take years to recover. Lakes become ponds, rivers become creeks. The water evaporates but any pollutants remain, becoming more concentrated in smaller amounts of water. If fish and water plants die, so do the birds and land animals that feed on them. Clouds of dust are stirred up by the winds, blackening the skies and polluting the air, suffocating livestock, even stripping the paint off buildings and cars.

With prolonged drought, farmers may not be able to earn a living. People move away to search for jobs. Rural towns close down once there are not enough children to keep a school open, no customers for local stores and businesses.

Finally, if rainfall is consistently low for long periods, the ecosystem itself changes, turning into desert. The water table drops below the point where not even a few seasons of rain will replenish it. The fragile, thin layer of topsoil is blown away, even deep-rooted grasses and trees cannot find water, and the people and animals leave for good.

OCEANS WARM

Though land areas are warming more than the oceans, the seas are warming, too. The water warms at the surface and gradually mixes into deeper water, with the warming becoming smaller the deeper you go. The average warming has been .037°C (.066°F) in the past fifty years.[6] It doesn't seem like much, but that's to a depth of 3,000 meters (10,000 feet). It takes a huge amount of heat energy to warm such a large volume of water.

Warmer sea temperatures affect first the plants and animals that live in the water. In general there are more fish in warmer waters, and they grow larger faster. But many sea creatures are extremely sensitive to temperature, and warming oceans can quickly change the location of the world's commercial fisheries. Cod, for example, will migrate to warmer waters; salmon prefer cooler.

Warming sea temperatures also threaten the world's coral reefs, home to 9 million different kinds of marine plants and animals. Coral reefs are among the oldest and richest ecosystems on the planet, but they are extremely fragile. Though they look like underwater gardens, they are actually made up of primitive animals called polyps. The polyps contain algae, which they need to live.

A number of things cause coral to expel their algae, including disease, shade, damage from ultraviolet radiation, changes in salinity and water temperature. If the water temperature rises just a degree or two, the polyps will expel their algae, leaving nothing but a white skeleton of calcium carbonate. If they do not replace their algae within weeks, the bleaching becomes permanent, and the corals die.

Warmer ocean waters also contribute to rising sea levels, because water expands with warmth, taking up more room.

Rising sea levels affect the world's coastlines in many ways. Shorelines and beaches are eroded. Salty sea water seeps into low-lying delta areas, many of which are heavily populated and used for intensive farming. During storms, waves reach areas that were once too far inland to be affected.

Rising water levels have a dramatic effect on all species living by the sea, from nesting shore birds to humans. One-third of the world's population lives within 100 kilometers (60 miles) of the sea, and more than

Carbonated Oceans

Increased carbon dioxide emissions are doing more than changing ocean temperatures. Scientists have recently discovered that the oceans have been pulling so much carbon dioxide out of the atmosphere so quickly that the sea's chemistry is changing. Oceans normally dilute the extra carbon dioxide in surface waters by mixing it into deep waters through natural circulation, but this is no longer happening fast enough. Instead, the carbon dioxide is turning into carbonic acid in the water.

Acidic oceans weaken coral skeletons and make them more susceptible to storm damage. The acid also corrodes the shells of shellfish and certain plankton, threatening the marine food chain.[7]

half of the world's twenty largest cities are on the coast. (A one-meter rise in sea level would be enough to flood most of New York City, including the subway system and all three major airports.[8])

SNOW AND ICE MELT
When average global temperatures rise only a degree or two, the greatest and earliest warming takes place in the cold polar and mountain regions. As ice sheets shrink,

there is less snow and ice to reflect sunlight back into space. Instead, the exposed land and dark oceans absorb the sun's rays, warming the atmosphere even more. Glaciers shrink, reducing the meltwater feeding rivers that provide fresh water to cities far below. In some cases, as glaciers melt faster than normal, glacial lakes fill up, putting pressure on their fragile banks (often made up of piles of rock and debris left behind by the retreating glacier). When the banks burst, floods and landslides head downhill.

As sea ice melts, new sources of oil and minerals become accessible to drilling, and new shipping lanes open up. But the exposed shorelines are also more vulnerable to storm waves and erosion. Melting sea ice, such as the Arctic ice cap, does not contribute to rising sea levels, since the melted water takes up the same amount of room as the floating ice. However, melting ice sheets on land do release large amounts of fresh, cold water into the sea, which contributes to sea level rise and affects the ocean currents. And although the continental ice sheets on Greenland and Antarctica are still very thick and cold in the middle, once large masses of ice begin to thin and melt, the melting is difficult to stop. In Greenland, large glaciers are already thinning around the edges and flowing into the sea more quickly as meltwater seeps through cracks to the rock beneath, making the rock slippery and accelerating the slide.[9] (The loss of polar ice may be counteracted by the increased precipitation that global warming will bring, but nobody knows for sure what will happen, especially in Antarctica.)

When permafrost (ground that is frozen year round) melts, once-solid ground turns into bog and becomes soft and unstable. Roads and runways crack, heave and sink. Buildings and pipelines shift and tilt on their foun-

Buried Treasure

Global warming is making one group of scientists happy — archeologists. As the world's glaciers and permafrost thaw, they are releasing tools, weapons, clothing and mummies that have been preserved in the ice for thousands of years. In the Italian Alps, the 5,300-year-old Iceman, the oldest complete human mummy ever found, was discovered poking out of a puddle on a melting glacier. In eastern Siberia, the mummy of the now-extinct woolly mammoth has been uncovered, and some optimistic researchers are even hoping to extract enough DNA to clone it.

Mummies can provide valuable information about how ancient people lived and died, what they ate, which illnesses they had. But some scientists have also suggested that melting ice sheets could unleash long-frozen viruses that are harmful to humans. Researchers recently extracted live virus from buried, frozen victims of the 1918 Spanish flu, in the hopes of studying it and perhaps creating a vaccine that might prevent another pandemic like the one that killed as many as 40 million people eighty years ago.

dations. Heavy trucks and machinery are no longer able to roll over frozen ground to reach isolated mining sites. Pools of meltwater sit stagnant in the summer, creating breeding grounds for mosquitoes.

Thawing permafrost also releases into the atmosphere carbon dioxide and methane — two major greenhouse gases long stored in the tundra's frozen peat.

FRESH WATER SUPPLIES ARE THREATENED

Shifting precipitation patterns, increased evaporation and shrinking glaciers all affect the amount and location of the world's fresh water — the water we all use to drink, wash and irrigate our food crops. But global warming also affects the quality of the water. Algae and microbes like warm water. Shallow lakes and rivers are more easily polluted than large, deep ones. Heavy rainfall and warm temperatures promote water-borne parasites that contaminate drinking water. Rising sea levels allow salt water to seep into freshwater coastal marshes and underground aquifers. A sudden spring melt can cause lakes and rivers to burst their banks, causing floods in valleys below, eroding mountainsides and draining freshwater reservoirs.

When the world's freshwater supplies are threatened, important questions arise. What happens when a mountain glacier shrinks until it can no longer supply drinking water to the millions of people who live downstream (half the people in the world use mountain water to drink, grow food and run their power plants)? Who owns the world's fresh water? Where do people who don't have fresh water go?

HABITATS SHIFT

Global warming should be good news for plants. Carbon dioxide stimulates photosynthesis, speeding growth. Longer growing seasons and more precipitation should mean bigger yields — especially for heat-loving plants such as soybeans, grapes and tomatoes — and perhaps allow multiple harvests in a single season. New growing zones should open up, creating new forests and new carbon sinks to absorb atmospheric carbon dioxide.

But shifting vegetation belts can be a tricky business. Plants may spread their seeds into warmer zones, but rainfall may be sparse, or soils may be too thin or lacking in nutrients to support the new vegetation (deciduous trees need deeper soils than shallow-rooted conifers, for example). As areas warm, existing cool-weather species may become stressed by the extra heat and unable to fight off disease.

As well, growing belts do not shift in tidy, smooth lines, or at an even pace. Warm-weather species may creep north during a spell of milder winters, only to be zapped by an unusually harsh freeze. Fall-seeded food crops may survive milder winters, but so may weeds and insects.

Some plants should become thicker and hardier in a

warmer climate, though some of the world's most important food crops may not. Rice fertility drops 10 percent for each Celsius degree above 30°C (86°F) during the plant's flowering.[10] Five consecutive days of temperatures exceeding 35°C (95°F) can ruin a corn crop.[11] Potatoes do badly when temperatures rise above 28°C (84°F).[12] Apples and other fruit trees need winter chill to form buds.

And no one knows how livestock will adapt to the stresses of climate change, and what the consequences for beef, pork and lamb producers will be.

When vegetation belts shift, the wild animals that feed on the plants try to follow. But some are better travelers than others. Birds' migration patterns are disrupted when their traditional resting and feeding stops disappear. Small, mobile animals are often the most adaptable (which is one reason why small mammals outlived dinosaurs during the mass extinction 65 million years ago). Crows, rats and flies, for instance, can thrive and reproduce almost anywhere.

Others will not be so lucky. As the world warms, forest fires, droughts, insect infestations and storms can destroy traditional habitats and force predators to move into neighboring ranges. These migrations can lead to new competitions for resources, changes in predation, even extinction for some species. The first and most dramatic impact would be on plants and creatures that already have limited ranges or a narrow supply of food sources, such as those that live in the mountains, polar regions or delicate marine ecosystems.

Because the Arctic ice is breaking up earlier in the summer, for example, polar bears, which hunt seals on the ice, have less time to put on enough weight to survive the winter and feed their young. A hunting season shortened by just one week can cause a polar bear to be 10 kilograms (22 pounds) lighter.[14] Researchers note that it is cubs and old bears that are dying in the greatest numbers, which is what happens to many species, including humans, in times of food shortages.

And what about the effect of shifting habitats on human populations?

In the past, climate change has often forced groups of people to move in search of food and livelihood. But with more than 6 billion people on the earth, these migrations are no longer simply a matter of packing up one's tent and following the caribou herd. Around the world, people have moved from rural areas to cities (by 2030, it is estimated that 60 percent of the world's population will be living in cities[15]), away from places that are too hot, too wet, too dry. Many major urban areas are so crowded that settlements have spread out to less desirable land — along low-lying coastlines that are particularly vulnerable to ocean storms, or flood-prone deltas.

But if the effects of climate change — heat stress, floods, droughts, insect-borne diseases, water shortages, shifting food sources — become chronic, many, many people will be on the move. Those who can hunker down in air-conditioned buildings, who can afford to repair their storm- or flood-damaged homes and buy clean water will probably be able to stay put. But for the hundreds of millions in the world who are already struggling, living from one harvest to the next, in countries with limited infrastructure and unstable governments, the migration will be desperate and massive.

Where will these people go? A 2003 report commissioned by the US Pentagon concluded that climate change could create a security risk for the United States as early as 2020.[16] With its large area, diverse climates, technological resources and wealth, the US would be better able to cope than most, but it may decide to defend its borders and coastline to keep out those fleeing their own countries.

The same thing could happen in other parts of the

world. Central Europe, for instance, would become crowded with people fleeing from a thawing Scandinavia in the north and a drought-prone Mediterranean in the south.[17]

And all over the globe there would be a battle for the world's arable land, energy and, especially, fresh water.

Chapter 5
The Frightening Numbers

> Prediction is very difficult, especially about the future.
>
> — Niels Bohr[1]

Scientists can tell us what the general effects of global warming will be, partly because of what has happened during warm periods in the past, and partly because the effects of the current warming can already be seen and measured in things like retreating glaciers and shifting habitats.

But scientists cannot tell us with absolute certainty the things many young people really want to know. How will climate change affect my house, my family, my community, my livelihood? Which will be the "good" places to live when it is time to get a job, buy a house, raise a family? Exactly when will these changes take place? In my lifetime? Or not until my children's generation?

One of the biggest problems with proposing action

on climate change is that scientists are often unable to make specific predictions. And yet they are constantly asked to do so. Governments want hard facts before they introduce unpopular legislation that will force reductions in greenhouse gas emissions. Insurance companies want to know how much they may have to pay out for climate-triggered disasters. (Insurance claims from hurricane damage in the US in 2004 and 2005, for example, came to $90 billion.[2]) Investors and consumers want to know where to put their money. (Invest in wind energy or the Scandinavian wine industry? Spend extra for a hybrid car? Build a house on a hillside or by the shore?) And the media looks for short, attention-grabbing statements that will fit into a newspaper headline.

So climate centers do issue predictions (or "projections," as they prefer to call them), based on computer models called General Circulation Models (GCMs). Researchers feed pieces of information into powerful supercomputers (the newest can perform 72 billion operations a second) — data on the amount of greenhouse gases in the atmosphere, the location of land and water, soil depth, the earth's rotation and orbit and many other factors. Sophisticated computer programs then apply the laws of physics to simulate the interactions between land, water, sunlight, air, ice, etc., and determine what the climate will be like over the next months, years, decades and even centuries.

Computer models are far from perfect — critics say

their results are so broad as to be next to useless — because no matter how sophisticated the model, it can never factor in the many, many complexities that affect climate. A tiny physical, chemical or biological action can cause turbulence in the air or water that may trigger major reactions somewhere else. This notion that a small random change can affect a larger outcome is called chaos theory, and climate is a prime example of how it works, because it is affected by millions of random events. This is the phenomenon that allows it to be dry over your house, but raining over your friend's house two blocks down the street.

Climate models also cannot resolve all the small-scale processes and features that drive day-to-day weather and, ultimately, climate. They cannot, for example, properly take into account the shape, height, thickness and size of clouds or predict what kinds of clouds there will be. Clouds are just too complicated. Some produce precipitation, some do not. Some block incoming solar radiation, some block outgoing heat. Climate models also cannot represent the effect of every heat-absorbing town or city, every valley or mountain.

As well, climate scientists can't say exactly when or how warming may be offset by human forces (such as technological advances or changes in practice that result in reduced emissions) or natural forces (such as solar or volcanic activity). It is extremely difficult, for instance, to estimate how much carbon dioxide we will continue to produce — something that depends on how much our population and economies grow, whether we cut or plant forests, whether we develop alternative energy sources or invent a new way to make energy.

Today, atmospheric carbon dioxide levels are about 380 ppm. Even if we stabilize or reduce greenhouse gas emissions, the level would still rise to 450 ppm by 2100, because of the carbon dioxide already in the atmosphere and because carbon sinks are not keeping up with current emissions. If we *continue* to increase greenhouse gas emissions at the present rate (called the "business as usual" scenario), atmospheric carbon dioxide levels will double over the next fifty to one hundred years.[3]

And if emissions increase even faster,[4] the news will be much worse.

As carbon dioxide levels rise, here are a few things climate models are predicting:

Average Global Temperatures Will Rise

All climate models say that temperatures will continue to rise. The most recent climate models predict that average global temperatures will rise by 3°C (5.5°F) by 2100,[5] a *rate* of warming that has not occurred in at least 10,000 years.

Not every place in the world will become just a little bit warmer at the same time. Almost all land areas will likely warm more than the predicted average, and the biggest warming will occur in the northern polar regions.

The changes will also not happen evenly or steadily. Despite the fact that 2004 was the fifth-warmest year on record, for example, in the northern hemisphere the spring of 2004 was the second-coldest.[6]

Sea Levels Will Rise

The IPCC predicts that average sea levels will rise by about 39.5 cm (15.5 inches) by 2100.[7] This is the average; the increase will vary around the world. Most of it will be due to the water warming and expanding, but about one-quarter of the rise will be caused by runoff from melting ice.[8]

No one predicts that the ice at the poles will melt completely, but with sufficient warming the West Antarctic Ice Shelf (WAIS) — a mass of ice propped up by island "pillars" — could become unstable enough to collapse and slide into the ocean, causing global sea levels to rise by several meters.

Snow and Ice Will Melt

As the Arctic and Greenland ice sheets melt, and with increased precipitation and river runoff, cold, fresh water will be released into the North Atlantic. A large enough influx would dilute the saltiness of the Gulf Stream. The current would become less salty and therefore less dense and no longer sink. Instead it would weaken and eventually stall, no longer bringing warm water from the southern Atlantic. The predictions vary, but some mod-

Alaska

Alaska is a well-studied area that provides a good example of what is already happening in the world's high latitudes as the planet warms up. The Columbia Glacier, for instance, a thick, moving river of ice 60 meters (200 feet) high and 5 kilometers (3 miles) wide, is receding at a rate of about half a mile a year.[9] It will disappear by the end of the century.

The average winter temperature in Alaska has increased 3°C (5.5°F) over the past fifty years — twice the rate of increase in the rest of the world.[10] Spring melt is taking place eight days earlier than it did in the 1920s. The beaver population is moving north, building dams that are polluting formerly clean waterways. Evergreen forests, stressed by longer, warmer summers and milder winters, are succumbing to infestations by insects such as the spruce bark beetle. Though annual precipitation is predicted to increase in the interior, the summers are already drier, so even though the growing season is now long enough and warm enough to support crops of barley, oats and hay, there is not enough rain for irrigation. In 2004, 2.5 million hectares (6.3 million acres) of forest — an area the size of New Hampshire — burned during the hottest and driest summer on record.[11]

As the permafrost thaws, pockets of ice trapped in soil are melting, causing the land above it to cave in. Eventually the thawed land will become solid ground, but in the meantime, electrical poles are falling down, roads and airport runways are buckling, and the foundations of the 3,200 kilometer (2,000-mile) Trans Alaska Pipeline are in jeopardy. (Alaska already spends $35 million a year repairing permafrost damage.[12])

els expect the Gulf Stream will weaken by 20 to 50 percent by 2100, and some say the current will stall entirely in two hundred to three hundred years.[13] One study reports that the ocean circulation in the North Atlantic has already weakened substantially.[14]

The results of a stalled Gulf Stream are uncertain, but many scientists think it could mean the collapse of the North Atlantic fisheries[15] and cooler weather in some of the most heavily populated and economically active areas of the world (Great Britain could have a climate similar to that of Labrador), affecting everything from shipping (rivers and harbors would freeze up earlier) to crops (which would be more vulnerable to frost) and heating costs.

Modelers also predict that half the mass of mountain glaciers and small ice caps will melt by 2100.

And at the Other End of the World...

Tuvalu is a small South Pacific nation made up of nine islands built on an ancient coral reef. Most of it sits less than 50 centimeters (20 inches) above sea level, and the island is expected to disappear by the end of the century. Many of its 11,000 citizens have already been evacuated to New Zealand. The country has begun legal action to gain compensation from greenhouse gas-producing countries, claiming these nations are responsible for the climate change that is causing their nation to be flooded.

Precipitation Patterns Will Change

Over all, annual precipitation will increase during the twenty-first century. But where will this precipitation fall, and when? Increased precipitation is predicted, for example, in Antarctica in winter, and in southeast Asia in summer. Australia, Central America and southern Africa, on the other hand, will receive less rainfall. The Mediterranean will receive less rain overall but more variable summer rainfall, increasing the likelihood of drought and flash floods. Increased drying during the summers in continental Africa and the North American prairies will reduce both the amount and quality of fresh water for irrigation, hydro power, water transportation and drinking.[16]

Vegetation Zones Will Shift

As climate zones shift, plant species will move into new growing areas, and the mix of species within areas will change. Most plants migrate at a rate of, at most, 1 kilometer per year,[17] but if the planet warms by only 2°C (3.5°F) over the next century, species will have to migrate seven times faster.[18] In intensively farmed mountain areas in countries near the equator, such as Ecuador and Rwanda, growing zones will move uphill.[19] In the mid latitudes, agricultural zones are expected to shift 200 to 300 kilometers (320 to 480 miles) for every degree Celsius of warming.[20]

For humans, one of the biggest concerns is the future

of the world's cereal crops. There will be a longer frost-free season, and therefore longer growing season, in Canada, Scandinavia, Iceland and Australia. In western China, models predict the farming season will be extended by as much as one month. But in the North American prairies, the Mediterranean, Latin America and southern Africa, warming will be accompanied by decreased precipitation, which may mean there is not enough water for crops.

Scientists predict that global warming will not likely affect the world's total food supply in this century. However, it *will* affect where food grows. Parts of the world already struggling, such as tropical and sub-tropical areas, will have fewer cereal crops. And if temperatures go up more than a few degrees Celsius, there will be an overall drop in agricultural activity, an increase in food prices around the world, as well as increased risk of famine in some developing countries. Warming of only 2°C (3.5°F) would, for example, make it too hot to grow coffee in Uganda — the base of the country's economy.[21]

Animals Will Migrate and Adapt — or Not

Some animals will not be able to migrate or adapt quickly enough to survive the effects of climate change. At the current rate of warming, the Great Barrier Reef will be wiped out within fifty years.[22] One study predicts that as a result of global warming, one-quarter of all land plants and animals — 1 million species — will either be extinct or becoming extinct by 2050.[23] It would be the largest mass extinction since the disappearance of the dinosaurs 65 million years ago.

And what about humans? No one predicts that the human race will be wiped out because of climate change. But global warming is expected to increase threats to human populations, especially those already on the margins. People who live in tropical countries will likely face poorer water and air quality. One study predicts that by 2050, 150 million people will be displaced because of global warming — 100 million due to sea level rise and flooding and 50 million because of drought.[24] Most models predict that disease-carrying species like mosquitoes and ticks will move with the warmer climate zones, bringing infectious diseases. Malaria already kills 2 million people each year, and 300 million to 500 million a year are affected. As the mosquito-borne disease moves into new territory, including Australia, North America and southern Europe, the local population's natural immunity will at first be very low, increasing the mortality rate.

• • •

The predictions are alarming. Maybe too alarming? When the latest IPCC report was released in 2001, some critics claimed that it painted an overly bleak picture of the future. But other recent studies have issued findings that are even more troubling.

Climate Change Around the World — Some Predictions[25]

Northwest Passage ice-free year round by 2050, reducing shipping journey from Asia to Europe by 7,000 km (4,300 miles)

Hudson Bay polar bears disappear by 2050

Great Lakes water levels 1 meter (3 feet) lower by 2050

Mississippi stream flow drops 30% by 2100

Sea level rise floods 1,800 sq km (700 sq miles) of Louisiana wetlands by 2050

Mild winters trigger infestations of mountain pine beetle, killing up to 80% of BC's mature lodgepole pine by 2020

Peru's Quelccaya Ice Cap, source of water for 10 million, gone by 2100

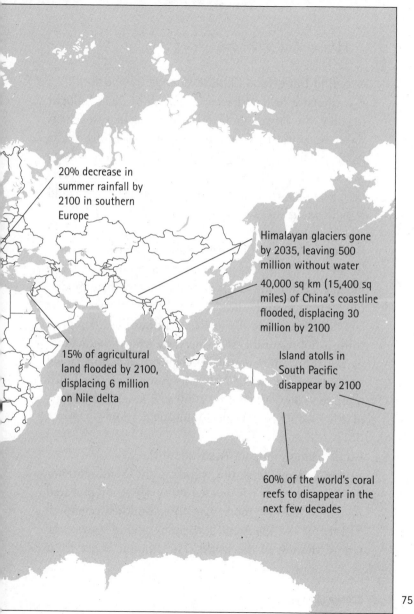

20% decrease in summer rainfall by 2100 in southern Europe

Himalayan glaciers gone by 2035, leaving 500 million without water

40,000 sq km (15,400 sq miles) of China's coastline flooded, displacing 30 million by 2100

15% of agricultural land flooded by 2100, displacing 6 million on Nile delta

Island atolls in South Pacific disappear by 2100

60% of the world's coral reefs to disappear in the next few decades

Here are just a few examples:

Avoiding Dangerous Climate Change Symposium[26]

A conference held in Exeter, England, in 2005 brought together the largest group of climate change scientists since the 2001 IPCC report. Although the 2001 IPCC report predicted that the West Antarctic Ice Sheet (WAIS) was so cold that it would not melt for centuries, experts in Exeter reported early signs that the ice sheet could already be disintegrating.

Hadley Centre for Climate Prediction and Research[27]

In 2001 the British government commissioned the Hadley Centre (one of the world's main climate modeling centers) to study the effects of climate change and its consequences, especially for Great Britain. The complete study is not due until 2007, but early reports predict that by the 2040s more than half of European summers will be warmer than that of 2003 (the year a heat wave killed more than 30,000), and by 2060 a summer like the one of 2003 would be considered unusually cool.

Arctic Climate Impact Assessment[28]

At the end of 2004, the world's eight Arctic nations, including the US, released a 1,200-page study prepared over four years by more than three hundred scientists. The report says the Arctic is warming at almost twice the rate of the rest of the world. The amount of sea ice is shrinking and the Greenland ice sheet is melting. The report predicts that in the next hundred years, Arctic temperatures will rise 4° to 7°C (7° to 12.5°F), and the permafrost line will retreat 300 kilometers (480 miles) north.

More recent studies reveal that both the Greenland ice sheet and the West Antarctic Ice Sheet are melting much more quickly than previously thought. Though the IPCC predicted in 2001 that global sea levels would rise no more than a meter (3 feet) by the end of the century, new research predicts melting ice from Greenland and Antarctica will cause sea levels to rise several meters by 2100.[29]

• • •

Nobody knows for sure precisely how high carbon dioxide levels will rise, or what exactly will happen as a result. But climatologists are certain about one thing. Although some people will initially benefit from climate change (some industrialized countries, for example, will enjoy longer growing seasons), in general, global warming will bring more harm than good to humans.[30] The problem is not just the fact of global warming — something the planet has experienced many times before. It is the speed of the warming, and the impact this will have on a crowded planet.

Chapter 6
The Tough Questions

> There is no other planet to which we can turn for help, or to which we can export our problems. Instead we need to learn to live within our means.
>
> — Jared Diamond[1]

The earth's atmosphere is becoming warmer. This warming is happening unusually quickly because humans are adding greenhouse gases, especially carbon dioxide, to the air. The effects of this accelerated warming will be at best expensive and disruptive. At worst they will bring life-threatening catastrophe, especially for the poorest people of the world.

It is too late to avoid all the harmful effects of global warming. But we *can* try to slow down the advance — to give ecosystems time to adapt, to reduce the threat to the world's fresh water and food supplies, and to prepare people of all nations to cope with the changes as best as they can.

It makes sense, then, to reduce the amount of carbon dioxide we are putting into the atmosphere.

There are two obvious ways to do this:

Reduce the amount of fossil fuels we burn. We will *have* to stop using these fuels eventually anyway, because there is a limited supply of gas, coal and, especially, oil underground.

It is ridiculous to think we can simply stop using energy, but we can reduce our dependence on fossil fuels by using energy less wastefully and more efficiently, and by replacing fossil fuels with other sources of energy. We need to find other ways to run our vehicles and machines and heat our buildings. Right now fossil fuels supply more than 80 percent of the world's energy, but the technology exists for us to obtain much more from renewable sources such as wind, solar, geothermal and biomass (fuels made from plants and other organic material).

Preserve and manage the world's existing forests, and plant new forests. Forests are carbon sinks, so by maintaining healthy forests, we can actually remove some of the carbon dioxide that is building up in the atmosphere.

This is the logical response to the problem of global warming.

So what is standing in the way?

The End of Oil — The Good News

There is a limited supply of oil buried underground, and half of it has been used up. Some experts say the supply has already peaked; others say there is enough for another forty years. But everyone agrees that most of the cheap, easy-to-get-at oil (the stuff that gushes out of the ground when you drill into it) is already gone. Though energy companies continue to squeeze and flush oil out of sand, shale and ocean bottoms, where there is still oil in large amounts, these methods are expensive. It costs trillions of dollars to develop new reserves and requires large amounts of energy just to get the oil out of the ground, transport it and refine it.

The dwindling supply of oil — as well as many other economic and political factors — is causing oil prices to rise, and as they go up, everything costs more, from gasoline and heating oil to plane tickets and any product that needs to be transported. Imported food will cost more (in North America, the average head of lettuce travels more than 4,000 kilometers/2,500 miles before it reaches our plates[2]), as will anything that contains oil products (from synthetic fabrics and cosmetics to asphalt and plastics) or that is made using oil-produced electricity.

But when oil prices go up and stay up, people do change their habits. They fly less, drive less, turn down their thermostats and install energy-efficient furnaces. They turn their attention to wind and solar power and other alternative energy sources. Between 1973 and 1986, when the price of oil was especially high in the US, more people bought fuel-efficient cars, speed limits were lowered, and the government gave tax credits to companies that developed sources of alternative energy.

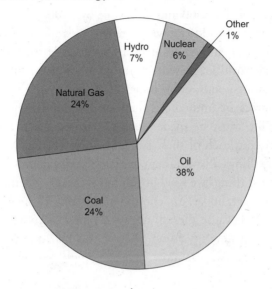

We use a *lot* of energy, and our energy needs are growing quickly. This is partly because the world's population is so big — 6.5 billion people and growing by 219,000 a day.[4] Practices such as cutting down a dozen trees to build one house, clearing a field to keep a cow, burning coal to heat a room or cook a meal — even using a car to take one person to work — are no longer practical. More people, more homes, more cars and more industry equal more carbon dioxide emissions. Which means the problem is growing faster than we can keep up with it, because even while some try to reduce their energy consumption, others are using more and more. (Electricity generation, for instance, is expected to double by 2020, quadruple by 2060 and quintuple by 2100.[5])

But the world's growing energy needs are due to more than an expanding population. Societies of the industrialized world are based on producing and buying more and more goods, even though it is clear that the planet's resources are limited. All the things that make up everyday life must be made, transported, used and thrown away, and each of these stages requires energy. We consume things, discard them, replace them, expand them — not just things we need but things we want, whether it is a new coat, house, car, computer or stereo. There is never enough because there is always a better or more stylish model, a new invention. Where you had one you can now have two, or more — cars, homes, phones, televisions.

Western culture, rapidly spreading throughout the globe, is based on acquiring stuff, and success is judged by whether one has more or better stuff than others. This is how economies grow and people and nations become rich. The whole idea of using less or producing less is unacceptable, so consumers are constantly encouraged to want more, buy more, use more.

Take the way we think about cars.

Next to electricity generation, one of the largest and fastest-growing sources of carbon dioxide emissions is road vehicles. Cars and trucks produce 30 percent of the world's carbon dioxide emissions. There are more than

500 million cars on the road today, and millions of new vehicles are manufactured each year.[6] (If every family in India and China owned a car, there would be one billion more cars on the roads.[7])

Early on we learn that cars are more than just a way to get from place to place. Cars symbolize freedom and status. People can see right away what kind of car you drive, and that tells them what kind of person you are. Are you rich and cool or geeky and shabby? Your car says it all.

Cars are so expensive that most people have to borrow money to buy one. (The average Canadian spends more money owning and operating a car than on housing, food or education.[8]) Yet many families in industrialized countries have two or more. Most are driven with just one person in them. We buy new cars even when the old ones still work, because manufacturers keep coming up with new colors, styles and accessories — cars with CD players and phones, cars with air-conditioners and seat warmers, even cars with built-in TVs, mini fridges and navigation systems.

Most cars can already go about as fast as highways will bear, so car companies make models that are more powerful, heavier and bigger, with more accessories. Things like air-conditioners, power windows and seat warmers add to fuel consumption by making cars heavier or by using more electricity. (The average vehicle contains forty pounds of wiring alone, needed to connect all of the car's electrical components.[9]) Automobile manufacturers know that heavy, powerful cars contribute directly to the rising level of greenhouse gases because they use more fuel, but they make them because people want to buy them. SUVs, which have been called one of the fastest-growing causes of global warming,[10] burn 45 percent more fuel than regular cars (even though they are mostly used for everyday driving in cities), but they have bigger profit margins because they are basically fancy pick-up trucks.

Cars even determine where and how we live. Modern suburbs are built for people who have cars. Shopping malls are surrounded by parking lots. There are drive-through restaurants, banks and libraries. Multi-lane highways, tunnels, bridges, overpasses and parking lots are built and then widened to accommodate growing amounts of traffic. (Two-thirds of the land in downtown Los Angeles is used for streets, highways and parking.[11])

While people in industrialized countries drive their cars to their homes in the suburbs, people in the developing world look on. Almost half of the world's people still live near or below the international poverty line of two US dollars per day. Yet through television and advertising they can see what they are missing, and they want to have the same thing.

They say they should not be the ones to suffer because the climate is changing. It's not the developing world that created global warming by burning huge amounts of fossil fuels in the first place. Surely the rich countries (20 percent of the world's population using 80 percent of the world's resources[12]) should pay for the damage they have caused, as well as reducing their own

China

China has a population of 1.3 billion people, four times more than the United States. Although almost half the population still lives below the poverty line,[13] in the past twenty years the average income in China has tripled, and so has the country's energy use.[14] The country is producing and buying more goods, and it is now the second-largest economy after the US.

China now has 24 million cars on the road,[15] and the country is making and selling a quarter of a million new cars each month.[16] If China had as many cars per capita as the United States (135 million in 2002), there would be 600 million cars in the country — more than all the cars in the world today.[17]

At a time when some countries are trying to shut down their coal-fired power plants (because coal is the most polluting fuel for carbon dioxide), China (which has vast supplies of coal) is planning to build more than five hundred coal-fired plants by 2030.[18] China is now the biggest consumer of coal and the second-biggest user of oil after the US, and many experts say that with its urgent and massive energy needs (expected to double by 2020), it will overtake the US as the largest producer of greenhouse gases by 2025.

energy use, while the developing countries catch up.

But the idea of reducing greenhouse gas emissions is, for many, simply unacceptable. Powerful forces — governments and corporations — actively campaign against climate change action by trying to dispute the science and by playing on the fears of people who don't want to put their good life at risk. (The biggest and most profitable company in the world, ExxonMobil, refuses to invest in clean technologies and instead pressures governments to resist action on global warming.) Governments in industrialized nations want to avoid the hard political decisions required to reduce carbon dioxide emissions. Corporations fear that reducing the use of fossil fuels will mean reduced profits. Meanwhile, people in the developing world want their own economies to grow, which means using more energy.

It's true that people are gradually using more energy from renewable energy sources, but so far these industries are far too small to fill the world's needs, and each one has its drawbacks. Wind and solar power cannot be stored and must be backed up by other sources. And the initial setup costs are expensive. (It takes a lot of money and energy, for example, to make the photovoltaic panels needed to produce solar electric power.) The International Energy Agency predicts that by 2030 renewables (excluding hydro) will contribute only 4 percent of the world's energy needs.[19] Even nuclear power cannot fill the gap.

The Problem with Nuclear Power

Nuclear energy is created by splitting uranium atoms to create heat. The heat is used to boil water and make steam to turn large turbines and produce electricity. Nuclear energy doesn't produce greenhouse gases (although fossil fuels are often used to extract and process the uranium, and to build and run the power plants), so some say it is the only realistic option for producing large amounts of carbon-free power, at least in the short term.

But the problems with nuclear energy are far from short term. To replace even one-third of the carbon dioxide emissions presently emitted by fossil fuel plants, the world would need to build 3,200 medium-sized nuclear plants.[20] These power plants would require heavy security, since the uranium and plutonium used in them can also be used to make nuclear weapons. As well, the spent fuel rods that hold the uranium contain dangerous levels of radioactivity that lasts for hundreds of thousands of years and cannot be disposed of safely.

The hazards of nuclear power came dramatically to the world's attention in 1986, when an accidental explosion at the Chernobyl nuclear power plant in Russia blasted radioactive material (equivalent to ten atomic bombs of the kind dropped on Hiroshima) 1.3 kilometers (4,000 feet) into the air. The fallout spread as far away as Italy and Sweden, poisoning crops, livestock and people.

As well, renewable energy cannot be competitive unless consumers pay the true cost of burning fossil fuels. Governments often subsidize the fossil fuel industry — by providing money to help companies discover and develop new oil fields, by giving oil companies tax breaks. Subsidies mask the true price of the gasoline we buy at the pumps, at times making gas in the United States, for example, less expensive than bottled water. If people paid the true costs of transporting oil, cleaning up oil spills, the military costs of protecting access to oil, the cost to public health of smog and pollution produced by tailpipe emissions, fossil fuels would be far more expensive. Consumers would balk at paying so much of their incomes just to keep themselves warm and their appliances running, individuals and industries would use energy more efficiently, and the prospect of, say, using wind power or buying a hybrid car would be considerably more attractive.

Compared to reducing our dependency on fossil fuels, planting and preserving forests may seem to be an easy way to tackle global warming. Yet this, too, is not as simple as it sounds. Without careful supervision and management, only a fraction of replanted trees survive, so many more trees must be planted to replace those that have been cut. As well, it is difficult to recapture areas that were once forest because they have been used for farmland and cities. Sometimes the soils have been depleted and are no longer suitable for trees.

Meanwhile, our consumption of wood continues to grow. (In spite of the "paperless" promise of the internet, for example, we are using more paper than ever.[21])

Tropical forests are especially difficult to preserve and replace, because they are such complicated ecosystems and because the people of developing countries have a growing need for cooking fuel, shelter and farmland. So much of the Amazon rainforest is being cleared for timber, cattle ranches and soybean farming that 40 percent of the remaining forest (the Amazon contains more than half of the world's existing rainforests) is expected to be gone by 2050.[22] Besides, developing countries deeply resent rich nations telling them not to cut their trees, when the greenhouse gas problem was created by the industrialized world in the first place. As one Brazilian official pointed out, "We're not going to stay poor because the rest of the world wants to breathe."[23]

Chapter 7
Facing the Music

Doubt, of whatever kind, can be ended in action alone.

— Thomas Carlyle

As with any difficult problem, there comes a point when it's time to stop talking — and reading — about how and why we got into this situation, who is to blame, what the difficulties are and what might or might not happen in the future. There comes a time for action.

We have thrown our smartest scientists at the climate change issue, and they agree that global warming is real, it's quickly getting worse, and its severity is due to human actions. So it is human actions that must solve the problem. We will waste valuable time if we keep worrying and stalling.

It's time to face the music.

But where do we go from here? What's to be done?

LESSEN THE IMPACT

Adapt

Even if we stop burning all fossil fuels today, the warming trend will continue for several decades, as the greenhouse gases already in the atmosphere continue to do their work. Scientists estimate that the world would have to reduce its carbon dioxide emissions by 40 to 60 percent just to *stabilize* the amount of carbon dioxide in the atmosphere.

It is obviously too late to avoid the effects of climate change completely, so plans must be put in place to deal with the warming that is already occurring.

This includes stopping development along reclaimed swamp land or coastlines that may be affected by rising sea levels or floods. Building dikes in low-lying, flood-prone areas and storm sewers to deal with run-off during flash floods in paved cities. Removing undergrowth from forests vulnerable to drought-triggered forest fires. Putting in place emergency procedures to help the poor and elderly during heat waves. Establishing early-warning systems to alert communities to floods and extreme weather, as well as evacuation procedures. Measuring, forecasting and securing freshwater supplies and building water-treatment facilities. Tracking and anticipating the spread of infectious diseases. Setting aside wildlife preserves and wilderness corridors to allow animals to migrate.

All these measures are sensible ideas for improving

human health and safety anyway, and investing in prevention will be less costly than repairing the damage caused by floods, droughts and forest fires.

Reduce Emissions

Most companies and individuals in industrialized countries will not take voluntary action to reduce their carbon dioxide emissions. Many businesses insist that reducing the use of fossil fuels will mean the loss of jobs and lower profits. Most do not want to use less energy or pay more for clean energy.

It's true that some corporations are realizing it is in their best interests to use some of their profits to take action on climate change. Not just because industry, with its huge appetite for energy, is a big part of the global warming problem and must therefore be part of the solution. Businesses don't want to be left behind as fossil fuels are abandoned in favor of cleaner technologies. Some also see that reducing emissions can actually be profitable, and they know that government-enforced carbon reduction policies are bound to come. Dupont, a chemical company, has already reduced its greenhouse gas emissions to 65 percent below 1990 levels, beating its original 2010 target date; it aims to derive 10 percent of its energy from renewable sources by 2010.[1] General Electric, once a big user of coal, has pledged to reduce its carbon dioxide emissions and invest in clean technologies like renewable energy and hydrogen fuel cells, hoping that its own customers, such as electricity companies, will continue to buy GE products and follow its lead.[5] Equally important, companies like Dupont and GE lobby governments to address global warming issues.

Elsewhere, corporations from software manufacturers to banks are retrofitting their offices to make them more energy efficient.

Still, the changes are not happening fast enough. So some governments are finally starting to take the lead, by taxing things they want to discourage (such as gas-guzzling vehicles), by subsidizing or removing taxes from things they want to encourage (such as public transit or fuel-efficient cars), by charging "congestion" fees for energy and transport use, and by enforcing reductions for large carbon dioxide emitters such as energy producers and automakers.

Some governments, for example, are rethinking the many ways they subsidize oil companies, keeping the price of oil artificially low, whether it is by providing money to find and develop new sources of oil or by reducing taxes on gasoline and heating oil. Developing countries such as Indonesia, Malaysia and Thailand are learning that keeping fuel prices low discourages conservation and fuel efficiency. It also means less money for needs like health care and education, and less money to invest in renewable energy.

In the United States, twenty-eight states, tired of waiting for the federal government to enact climate change legislation, are taking their own action to reduce greenhouse gas emissions.[6] In Oregon, any new gas-fired power plant is required to be 17 percent more efficient than any other gas-fired plant in the entire country.[7] California has pledged to reduce its greenhouse gas emissions to 80 percent of 1990 levels by 2050.[8] Improved fuel-efficiency standards set by state legislators are forcing automakers to reduce greenhouse emissions by as much as 30 percent by 2014, creating a rising demand for hybrid vehicles that run partly on gas and partly on batteries. The battery is recharged as the engine runs, and the cars routinely use half as much gas as regular vehicles.

Elsewhere, the state of West Australia has formed a

partnership with an oil company, car company, fuel-cell manufacturer and university to fund three hydrogen fuel-cell powered buses in the capital city of Perth. And governments from China to Europe are subsidizing city buses that run on hydrogen, providing rebates or tax relief for people who buy hybrid cars, providing stable funding for buses, streetcars, trains and bike lanes.

Develop Renewable Energy Sources

Although it will be some time before wind, solar, geo-thermal, tidal and biomass can provide enough power to replace fossil fuels, the renewable energy field is growing. The countries of the European Union, for instance, plan to produce 22 percent of their electricity and 12 percent of all their energy from renewable sources by 2010.[9] Scotland plans to generate 40 percent of its electricity with renewables by 2020 with onshore wind farms, wind turbines built offshore on former oil rig platforms, and by capturing ocean wave energy.[10]

Wind power is the fastest-growing source of energy today. The generating capacity around the world has tripled in the past five years, with countries like Denmark, Germany and Spain taking the lead. Germany is on track to supply 25 percent of its electricity with wind by 2025. To help make wind power competitive, producers are paid above-market prices for the power they feed into the grid, and the costs are shared by all the electricity users in the country. As a result, between 1990 and 2000, the average cost of manufacturing wind turbines in Germany fell by 43 percent. By 2020 half the population of Europe is projected to get its residential electricity from wind.[11]

North America is far behind. It supplies less than 1 percent of its electricity with wind, even though it has lots of wide-open spaces and coastlines that are ideal for wind turbines. But things are changing, and some states, provinces and communities are taking the initiative in the hopes of reducing their dependency on fossil fuels. A wind farm in Prince Edward Island (a Canadian province of 140,000) now supplies 5 percent of the province's electricity, and the province plans to eventually generate all its electricity with wind.[12] Windblown island communities like PEI and Unst in the Shetland Isles are investing in wind/hydrogen operations (using wind to create hydrogen that will power fuel cells at times when the wind doesn't blow).

Other renewable energy industries are also making headway, sometimes with the help of oil companies that are hedging their bets by investing in other kinds of energy (BP, for example, is one of the top producers of photovoltaic panels). A power station on the Bay of Fundy in eastern Canada harnesses the energy of incoming and outgoing tides. The city of Stockholm is building biogas plants to turn sewage and organic waste into fuel. Elsewhere, the smart, organized people of Iceland are harnessing the steam from hot water that lies deep below

Wind Power

With wind power, wind turns the blades on tall windmills, or turbines, and the blades turn a shaft attached to a generator, which produces electricity. The electricity is then fed through underground transmission lines to the power grid. There are no greenhouse gas emissions when the energy is produced.

Many places in the world have open land and exposed coastlines ideal for wind turbines (wind installations can even be located offshore, poking out of the water). But there are barriers. Geographers must measure and test sites over long periods to see whether there is enough consistent wind to turn the turbines. Large tracts of land must be obtained, requiring the cooperation of the people who live in the area, some of whom may object to the noise and look of the windmills, and their effect on birds and other wildlife. And transmission lines must be built to carry the power, sometimes long distances, to where it will actually be used.

Finally, wind power cannot be stored like a tank of oil. Even in areas with lots of wind, there are still periods of calm (even ideal sites produce power only 30 to 40 percent of the time), and some form of backup energy must be used.

But the technology is constantly being improved. For example, engineers have developed smaller, more efficient turbines that rotate vertically (like a merry-go-round), allowing them to harvest more of the wind's available energy as well as being cheaper to manufacture and repair than the giant big-blade windmills.

their country to run turbines and produce hydrogen. Geothermal power, unlike wind and solar power, is constant, and the power is used to provide half of the country's energy.[13] In California, Nevada and Utah, geothermal power now produces as much energy as five nuclear reactors.[14]

Replacing fossil fuels with renewable energy may not be possible yet, but it will happen more quickly if governments help — by providing grants or loans for the construction of things like new wind turbines, and by requiring state-owned utility companies to purchase a percentage of their electricity from renewable sources. A study commissioned by Shell Oil claims that alternative energy sources could supply half the world's energy needs by 2050.[15]

Develop New Technologies

Human invention and progress created the global warming problem, and now we must hope our ingenuity will help solve it, the way the invention of the car, airplane, electric light and computers have transformed modern life in the past. Some are even counting on clever engineers and scientists to come up with a way to combat global warming that will not require any of us to change our present behavior.

The ideas range from the fantastic (launching gigantic mirrors into space to deflect sunlight away from the earth) to the mundane (making roads white to reflect light), but they are all being pursued at some level. We would not have to reduce our consumption of fossil fuels, for example, if we could stop the carbon dioxide from going into the atmosphere — by capturing the carbon dioxide and pumping it into the cold water at the bottom of the ocean (deep ocean sequestration), or into old underground oil reservoirs or coal beds. Some of these solutions have good potential; pumping carbon back into rocks is already being done in Algeria and Saskatchewan, Canada. Others have potentially dangerous consequences (the damaging effects of extra carbon dioxide on ocean chemistry, for example, have only recently been recognized).

We would also not have to reduce our energy consumption if we could find another way to generate power, which is why some are pinning their hopes on the hydrogen fuel cell.

Instead of running on fossil fuels, fuel cells combine hydrogen (H_2) and oxygen (O), causing an electrochemical reaction that produces water, heat and electricity. The oxygen is drawn from the air. The hydrogen can be obtained from a number of sources, from natural gas to plain water. Because there is no combustion, fuel cells produce no greenhouse gas emissions when they are used. But it does take energy to obtain the hydrogen in the first place. If the hydrogen is obtained from water (H_2O), for example, the hydrogen (H_2) and oxygen (O) molecules must be separated using an electric current (electrolysis) — a current that must be created by some outside source of energy.

Fuel cells can be used to power small items like laptops and cellphones, or they can be stacked to power factory generators or heat buildings. They can also be used to run cars.

In a fuel-cell car, instead of gasoline, hydrogen is stored in a pressurized tank, and the internal combustion engine is replaced by an electric motor powered by a fuel cell instead of a battery. Fuel cells in cars may one day even be used to run household appliances when the car is sitting in the garage. And the waste heat could be used to heat buildings or water.

Fuel cells have so much potential that governments, car manufacturers and energy companies are all investing in their development. Test buildings are being heated and powered by fuel cells — hotels in New Jersey, university residences in Toronto, homes in Belgium, businesses in Germany and Spain. Fuel cells are being used to run forklifts and delivery vehicles, and test cars are on the road in Vancouver, Orlando, Sacramento and Michigan.

But so far, the cost of making fuel cells is so high that heavy subsidies are required to attract customers. And there is a long way to go before they will be light, compact, powerful and durable enough to replace the internal combustion engine. As well, the hydrogen must be produced and stored in a compressed form, and a network of hydrogen refueling stations will have to be built.

Fuel-cell fans argue that this network will expand naturally, just as a network of gas stations grew up when the first automobiles hit the roads, but most experts admit it will be several years, perhaps decades, before fuel-cell cars are a common sight.

Still, all these new technologies and areas of research must be pursued. That's good news for today's students, because global warming is opening up a whole range of opportunities and jobs for young scientists and engineers in a variety of disciplines. Food engineers are developing drought-resistant and salt-tolerant cereal crops. Epidemiologists and biologists are learning how best to cope with warming-triggered diseases and pests. Biologists are tracking bird and butterfly migrations. Engineers are addressing the important task of preserving the planet's clean water, through conservation, efficient irrigation and desalination and purification techniques. Others are designing more energy-efficient cars, hybrid train locomotives that use half the fuel of traditional locomotives as well as reducing greenhouse gas emissions by up to 90 percent, and more efficient household appliances (in Japan you can buy a refrigerator that uses one-eighth the amount of energy as most ten-year-old fridges, and a buzzer sounds if you leave the door open for more than thirty seconds![16]).

Meanwhile, climatologists carry on with their work — observing climate, improving computer models, studying ice cores and measuring glaciers. And in all fields, scientists continue to review the work of their peers and question and revise what they know. Because science does not sit still, especially the science of climate change.

Plant Trees

Although human populations are growing and spreading out, there are still plenty of places in the world where we can plant trees—increasing the areas of forests that act as carbon sinks.

China, for instance, plans to plant 35 million hectares (86.5 million acres) of new forest ("a great green wall") by 2050.[17] New carbon sinks are also being created in the southern and eastern US as abandoned farmland and logged areas are set aside for new forests. Japan, which supports one of the world's highest population densities, has increased its forested areas to the point that 80 percent of the country is forested — the highest percentage of any industrialized nation.[18]

Meanwhile, botanical engineers are using pruning and planting techniques to develop fast-growing trees with denser root systems that will take in water and nutrients more efficiently. Other biologists are dispersing seeds to give trees a head start at expanding their range into new habitats.

SHARE THE BURDEN

Restraint for People in Rich Nations

Industrialized nations got rich by burning fossil fuels and creating the climate change problem in the first place. These countries continue to have higher per capita emissions of greenhouse gases than developing countries, because they are wealthy enough to consume wastefully. So shouldn't it be up to these rich nations to assume most of the burden of fixing the problem?

For individuals, this means exercising some restraint at home. Domestic consumption accounts for 12 percent of the total energy use in industrialized countries,[19] so people must make changes in their day-to-day lives to reduce the amount of energy they use. That means only using as much power as one really needs, and using it as efficiently as possible. It means considering the energy cost of everything we eat and use. Where and how was it made? How far did it come and how was it transported?

Some of the changes are pathetically simple and involve no hardship whatsoever. Choosing to heat up the lasagna in the microwave or toaster oven instead of turning on the big oven (a toaster oven uses half as much energy as a standard oven). Turning out the lights when no one is in the room. Turning down the thermostat in the winter and turning it up in the summer to save fuel in heating and air-conditioning. Turning off the computer at night (energy "vampires" such as TVs, modems, cordless phones and stereos suck power through their plugs even when they are turned off, making up 10 percent of a typical North American home's energy use).

Many of these choices can be easily extended to the workplace. Japan, for example, has launched a nationwide campaign to encourage office workers not to wear jackets and ties in hot weather, allowing them to use less air-conditioning.[20]

More people are thinking before they buy, choosing goods that are energy efficient and last longer, and buying locally grown foods that haven't been transported long distances. They are checking out energy-efficient appliances, battery-operated bikes and scooters and hybrid cars. (If everyone replaced their appliances with the most efficient models, electricity consumption would drop by more than half.[21]) Some are using public transportation more and driving less (a person driving alone in a car produces eight times as much carbon dioxide as someone taking public transit[22]), and considering whether big, heavy cars with lots of horsepower are really better than small, fuel-efficient ones. As the demand for public transit increases, the system becomes more efficient. Buses, streetcars and trains run more frequently, routes are expanded and more people are inclined to leave their cars behind.

Decisions like these reflect some hard thinking about what kinds of communities we want to live in. Neighborhoods with sidewalks, parks and easy access to

The Simple Solution

Even with the growing use of hybrid vehicles and low-emission fuels, one of the very simplest ways for individuals to reduce carbon dioxide emissions is to leave their cars behind. In addition to creating carbon dioxide pollution, congested roadways, smog and parking headaches have prompted cities around the world to encourage citizens not to drive by promoting bike racks on buses, bike lanes on main streets, toll lanes (where cars with just one person in them must pay an extra fee), commuter parking lots on city outskirts and car-sharing programs. Paris plans to prohibit all but commercial vehicles and local residents' cars from the city center. London levies congestion charges for those who drive in the city core. Other cities take special action during Car Free Days or Mobility Weeks — banning cars from the city core in Bogotá and Copenhagen, offering free public transit in Basel, sponsoring competitions to see who can drive their cars the least in Madison, Wisconsin.

public transit? Homes with solar water heaters, located where you can walk to school, work, the grocery store?

How much time and money do you want to spend driving? Do citizens have a right to drive cars? Should drivers of polluting cars have to pay more for parking? Should drivers be fined for idling their cars, which releases carbon dioxide into the atmosphere without even getting you anywhere? Should drive-through restaurants,

Making Your Voice Heard

It is sometimes hard to believe that a single person can have a say on a big issue like climate change. That's one reason why people turn to environmental groups.

Environmental groups push for action on issues such as climate change by informing the public and by pressuring governments and industry. The groups are supported by their members, and many have special membership rates for students. The largest groups are organized worldwide, and they are skilled at using the media and the internet. They may also lend their support to political parties that focus on environmental issues.

Environmental groups do not speak with one voice, and they have different ways of getting across their message. The World Wildlife Fund (whose patron is Britain's Prince Philip), might encourage citizens to send postcards urging government leaders to save the polar bears, whose habitat and feeding cycles are threatened by climate change.

Greenpeace, on the other hand, launches media-savvy campaigns that are often dramatic and angry, and part of their mandate is to carry out "peaceful acts of civil disobedience." Greenpeace, for instance, will charge the giant oil company ExxonMobil (which

where cars do little but idle, have to pay a penalty for encouraging pollution? Should they be banned altogether?

People are even questioning the need for air travel. Not only is carbon dioxide introduced into the atmosphere when fuel is burned (the airline industry emits recently reported the highest corporate profit in history) with "crimes against the planet." Protesters will display giant images of floods and dress up like Esso tigers outside the company's annual general meeting, telling shareholders about Exxon's shameful record on climate change.

Environmental groups can be a powerful voice. In 2003, a power company sought permission to build an underwater gas pipeline in British Columbia. Environmental groups convinced the Canadian government to insist that the company provide a plan to reduce the greenhouse gases that would be produced by burning the gas transported through the pipeline. In Great Britain a campaign (Stop Esso Day) stopped more than one million people from buying gas at Esso stations.

Consumers can use their buying and voting power to influence the activities of corporations and governments. If consumers don't buy SUVs, car companies will stop making them. If voters campaign for bike lanes, cities will build them. If customers flock to a laundry that uses solar panels to heat its water, other businesses will follow the example. And if voters turn their backs on politicians who ignore the threat of climate change, governments will change their policies.

twice as much carbon dioxide in a year as the entire country of India[23]); producing highly refined jet fuels uses vast amounts of energy, and cloud cover is affected by the water vapor in jet effluent trails.

Some energy-conservation tips may seem too small to make a difference. How can switching a lightbulb help fight global warming? Yet this is one of the easiest and most effective ways to reduce a family's carbon dioxide emissions. Lighting uses one-fifth of the electricity brought into the average home. Compact fluorescent lightbulbs use one-quarter to one-third as much electricity as incandescent bulbs, and they last up to ten times longer. And if every household in the United States changed all its lightbulbs, for example, the growth in US carbon dioxide emissions could be stopped.[24] Turning down the thermostat on the water heater (to 50°C/120°F) can save 540 kilograms (1,200 pounds) of carbon dioxide a year. Recycling a single aluminum pop can save enough energy to run a television for three hours.[25]

Besides, restraining our use of fossil fuels has other bonuses apart from fending off the worst effects of climate change — saving money, cleaner air, less traffic clogging cities, greener communities. And people who walk and bike instead of driving are healthier and less likely to be overweight than those who drive.

Smarter Progress for Poorer Nations

While rich nations restrain their own consumption, developing countries must catch up by improving their economies and lifestyles. But poorer nations can benefit from the knowledge rich countries have accumulated, and from the mistakes they have made.

Some developing countries, for instance, are already

far ahead in terms of efficient, widely used public transit, car sharing, road tolls for private vehicles, widespread use of compact fluorescent lightbulbs, solar hot water heaters, small vehicles and buses run on alternative fuels or electric batteries. Bogotá, Colombia, has car-free days, a cheap, reliable bus system and a large network of bike and pedestrian routes connecting outlying parts of the city with downtown.[26] Costa Rica invests heavily in wind energy and has vowed to phase out the use of all fossil fuels. China plans to build and refurbish 40,000 kilometers (25,000 miles) of its railroads by 2010.[27]

Developing countries may be in the best position to use new technologies, especially if industrialized nations will help them get set up, because their energy needs are still small, and many are not attached to a central power grid (one-third of the world's population has no access to electricity from a central source[28]). Solar cooking stoves and wind turbines to pump water or charge batteries, for example, would mean huge advances for the people of poorer nations. So in the Himalayas, communities are developing small-scale hydroelectric schemes to provide villages with electricity without creating greenhouse gas emissions. In many parts of the developing world, biofuels made from renewable sources such as sugar cane, pig manure, rice hulls and coconut shells are being used to generate power in rural areas, boosting the economies of small communities.

Climate-friendly progress for developing countries is

Good News Around the World[29]

Scotland builds Europe's largest wind farm

Iceland uses geothermal power to heat 87% of its buildings; aims to be free of fossil fuels by 2050

Maine forbids new building on shorelines vulnerable to rising sea levels

London drivers pay "congestion charges" to drive in the city core

California plans 50 to 100 hydrogen fueling stations by 2010

In Philadelphia white roof covers reduce top-floor temperatures of rooming houses by 3°C (5.5°F)

Los Angeles plans to build the world's largest solar power facility

Louisiana builds wave-dampening fences to protect Mississippi delta

Barcelona requires new houses to have solar water heaters

Chilean farmer develops technology to reduce methane emissions from hog manure and sells greenhouse gas emission credits to power companies in Canada and Japan

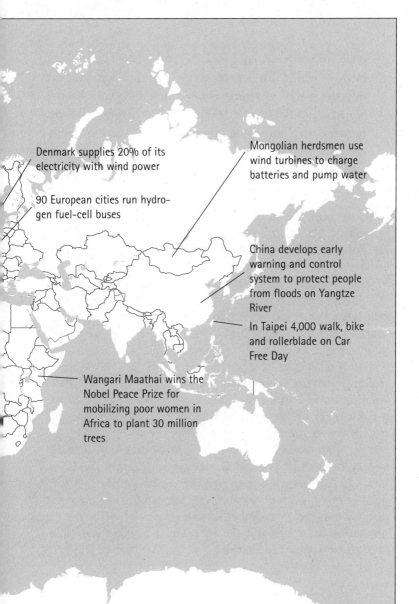

Denmark supplies 20% of its electricity with wind power

90 European cities run hydrogen fuel-cell buses

Mongolian herdsmen use wind turbines to charge batteries and pump water

China develops early warning and control system to protect people from floods on Yangtze River

In Taipei 4,000 walk, bike and rollerblade on Car Free Day

Wangari Maathai wins the Nobel Peace Prize for mobilizing poor women in Africa to plant 30 million trees good for everyone. Poor nations cannot be allowed to lag farther behind. As writer Jared Diamond has pointed out, the world's environmental trouble spots are also the world's political trouble spots; the bigger the gap between rich and poor, the more politically unstable societies become.[30]

Kyoto Protocol[31]

In February 2005, after more than seven years of painfully slow negotiations, the Kyoto Protocol, the first international framework to address climate change, was ratified.

According to the agreement, industrialized countries must reduce their overall emissions of greenhouse gases to (on average) 5.2 percent below 1990 levels by 2012. To meet their targets, industrialized countries can buy "credits" by helping other countries lower their greenhouse gas emissions. If, for example, a developing country uses renewable energy or renovates its fossil-fuel burning plants to reduce greenhouse gas emissions, that country can sell its credits to industrialized countries and use the money to boost its own economy. A country like Brazil could also get credits for replanting or maintaining its vast areas of forest.

The Kyoto Protocol is controversial, partly because it does not treat all countries equally. Large developing countries like China, India and Brazil do not have to limit their carbon dioxide emissions, despite their fast-growing economies. Some critics say the credit system allows the rich countries, who are responsible for most of the greenhouse gas emissions in the first place, to buy their way out of meeting their obligations, without changing their own environmen-

Climate change is such a large-scale problem that there is no magic bullet, no single solution. Instead we must pursue action on all fronts — adapt to the inevitable warming, pursue technological solutions, use energy wisely, reduce the gap between rich and poor nations, make polluters pay. This will require scientists, industry

tally destructive lifestyles. Many think the goals will be impossible to reach (many countries that initially agreed to cut have actually increased their emissions in recent years) and are too paltry to have a real effect on global warming anyway. Although one report suggests industrialized countries that signed onto Kyoto will make their overall target, worldwide, greenhouse gas emissions will increase by 50 percent between now and 2030.[32]

Kyoto may be just a tiny first step in the fight against global warming, but it does accomplish a number of important things. It puts climate change on the public agenda as a global issue that affects us all. It sets specific targets and deadlines that force people to get on with the task of reducing greenhouse gas emissions. It offers flexibility in how these targets can be reached. And it recognizes that rich countries must take the lead.

In December 2005, thousands of climate change experts and government representatives from around the world met in Montreal, Canada, where they agreed to discuss timelines and targets for the next stage of emissions reductions. They also agreed to pursue technologies such as carbon sequestration, and to develop strategies to adapt to the warming that is already occurring.

and governments at all levels to work together. And it will require all of us to sit up, pay attention and think about the consequences of our daily actions.

• • •

Some have compared the world's reaction to climate change to the people on the *Titanic*. The rich passengers in the first-class sections were busy eating fine meals, swimming in the ship's pool and exercising in the gymnasium. The poorer people on the lower decks were busy feeding their babies and worrying about starting new lives on another continent. Down in the boiler room, tired stokers shoveled coal into the ship's giant furnaces. And up in the cockpit, the captain and officers sat proudly, happy to be piloting the biggest, most modern, technologically advanced ship the world had ever seen.

Everyone was busy. Nobody saw the iceberg. Nobody was worried that there were not enough lifeboats.

Notes

1 Climate Change Is Here, and It's Real

1. Stephen Byers, co-chair, International Climate Change Taskforce, on the release of "Meeting the Climate Challenge," a report produced by the Institute for Public Policy Research (UK), the Centre for American Progress (US) and the Australia Institute, January 2005, www.ippr.org.uk/pressreleases/?id=1264.

2. Researchers have recently revised their estimates of heat-related deaths in Europe from June to August 2003 to 44,000 or more. Eurosurveillance, "The 2003 European Heat Waves," July 2005, www.eurosurveillance.org/em/v10n07/1007-222.asp.

3. Based on a chart provided by the Climatic Research Unit and the UK Met Office Hadley Centre, 2006, www.cru.uea.uk/cru/info/warming/.

4. US National Aeronautics and Space Administration (NASA) press release, January 24, 2006,www.nasa.gov/vision /earth/environment/2005_warmest.html. Two other organizations, the World Meteorological Organization (WMO) and the National Oceanic and Atmospheric Administration (NOAA) reported that 2005 had been the second-warmest year on record. NASA researchers claim those studies did not include certain data from the Arctic.

5. The size of the Larsen B ice shelf that disintegrated was 3,250 square kilometers (1,250 square miles), containing 500 billion metric tons of ice. About 87 percent of the glaciers on the Antarctic Peninsula (the finger of land that

sticks up just below South America) are retreating. A.J. Cook et al, "Retreating Glacial Fronts on the Antarctic Peninsula Over the Past Half-Century," *Science*, April 22, 2005.

6. Between 2000 and 2004 these lakes lost 50 percent of their volume. Bureau of Reclamation, Upper Colorado Region Water Operations, "Lake Power, Glen Canyon Dam – Current Status," June 6, 2005, www.usbr.gov/uc/water/crsp/cs/gcd.html.

7. Tim P. Barnett et al, "Penetration of Human-induced Warming into the World's Oceans," *Science*, July 8, 2005.

8. International Institute for Sustainable Development, "Sachs Harbour Observations on Climate Change," www.iisd.org/climate/arctic/sachs_harbour.asp.

9. International Institute for Sustainable Development, International Year of Mountains, "Mountain Heritage of Pakistan: The Essential Quest," 66. International Year of Mountains, www.mountains2002.org.

10. Timothy J. Osborn et al, "The Spatial Extent of 20th-Century Warming in the Context of the Past 1200 Years," *Science*, February 10, 2006.

11. J. R. Petit et al, "Vostok Ice Core Data for 420,000 Years," IGBP PAGES/World Data Center for Paleoclimatology Data Contribution series #2001-076. NOAA/NGDC Paleoclimatology Program, Boulder, 2001.

12. Edward J. Brook, "Tiny Bubbles Tell All," *Science*, November 25, 2005.

13. Leaf pores shut down when there is lots of carbon dioxide in the air. Because trees, like all plants, take in carbon dioxide to put on growth, rings in the trunk are thicker during years when carbon dioxide levels are higher. Bands on corals reflect seawater temperatures the same way tree rings reflect air temperature.

14. Michael Williams, *Deforesting the Earth* (Chicago: University of Chicago Press, 2003), 500; National Geographic Maps, *A World Transformed*, September 2002.

15. United Nations Framework Convention on Climate Change (NFCCC), United Nations Statistics Division, 2003, http://unstats.un.org/unsd/mi/.

16. Elizabeth Kolbert, "The Climate of Man - III," *New Yorker*, May 9, 2005.

17. Linda McQuaig, *It's the Crude, Dude: War, Big Oil, and the Fight for the Planet* (Toronto: Anchor, 2005), 333.

18. (Western Canada) In 2003, 250,000 hectares of forest burned down in British Columbia; 50,000 people were evacuated and damages totaled $550 million. Charles Anderson and Lori Culbert, *Wildfire: British Columbia Burns* (Vancouver: Greystone, 2003), 10. (Hudson Bay) Government of Canada, www.climatechange.gc.ca. (Atlantic beaches) Mark Lynas, *High Tide: The Truth About Our Climate Crisis*, (New York: Picador, 2004), 112. (US coast) Aristides Patrinos and Anjuli Bamzai, "Policy Needs Robust Climate Science," *Nature,* November 17, 2005. (Costa Rica) Alan Pounds et al, *Nature*, January 12, 2006. (Antarctica) J. Turner et al, "Significant Warming of the Antarctic Winter Troposphere, *Science*, March 31, 2006. (Perth) Tim Flannery, *The Weather Makers: How We Are Changing the Climate and What It Means for Life on Earth* (Toronto: HarperCollins, 2005), 129. (Nepal) United Nations Environment Programme (UNEP), www.rrcap. unep.org/issues/glof. (Fiji and Samoa) Lynas, 113. (Yellow River) Y. Ding et al, "Yellow River at Risk: An Assessment of the Impacts of Climate Change on the Yellow River Source Region," Greenpeace, 2005, http://activism. greenpeace.org/yellowriver/yrs_english_web.pdf. (China desert), *Ecosystems and Human Well-Being Desertification*

Synthesis. (Northern Hemisphere) Intergovernmental Panel on Climate Change (IPCC) Third Assessment Report, *Climate Change 2001, Synthesis Report* (Cambridge: Cambridge University Press, 2001), 6. (Britain) *National Geographic*, September 2004.

19. Thomas J. Crowley, "Causes of Climate Change Over the Past 1000 Years," *Science*, July 14, 2000.

2 How We Got Here

1. At a temperature of 33° to 42°C (91° to 107°F). Woods Hole Oceanographic Institute, "An Ocean Warmer Than a Hot Tub," *Oceanus*, February 17, 2006.

2. Time-Life Books, *The Natural World* (Alexandria, Virginia: Time-Life Books, 1991), 87.

3. Clive Ponting, *A Green History of the World: The Environment and the Collapse of Great Civilizations* (New York: Penguin, 1991), 255, 256.

4. An area of 12,075 to 19,823 square kilometers per year, including up to 1,200 square kilometers of conservation lands. Gregory P. Asner et al, "Selective Logging in the Brazilian Amazon," *Science*, October 21, 2005.

5. Ponting, 257.

6. The concrete industry is responsible for one-fifth of the carbon dioxide emissions produced by humans, Carbon Dioxide Information Analysis Center, www.cdiac.esd .ornl.gov/ndp030/global00.ems.

7. Ponting, 253.

8. United Nations, World Population Prospects, Population Reference Bureau, www.prb.org/content/Navigation Menu/PRB/Educators/Human_Population/Population _Growth/Population_Growth/htm.

9. National Geographic Maps, *A World Transformed*, September 2002.

3 How the Climate System Works

1. Yahai Lu and Ralf Conrad, "In Situ Stable Isotope Probing of Methanogenic Archaea in the Rice Rhizosphere," *Science*, August 12, 2005.

2. Michael Williams, *Deforesting the Earth* (Chicago: University of Chicago Press, 2003), 436.

3. Based on IPCC Second Assessment Report (SAR), *Climate Change 1995: Impacts. Adaptations and Mitigation of Climate Change* (Cambridge: Cambridge University Press, 1996), www.grida.no/climate/vital/32.htm.

4. *National Geographic*, May 1998; Canadian Centre for Policy Alternatives, *CCPA Monitor*, October 2004; J Jonathan Weiner, *The Next One Hundred Years: Shaping the Fate of Our Living Earth* (New York: Bantam, 1990), 182; Intergovernmental Oceanographic Commission, 2004; IPCC, *Climate Change 2001: The Scientific Basis* (Cambridge: Cambridge University Press, 2001), 188; John Houghton, *Global Warming: The Complete Briefing*, Third Edition (Cambridge: Cambridge University Press, 2004), 9.

4 The Effects of Climate Change

1. James Lovelock in a speech to the Canadian Nuclear Association, reported in *Globe and Mail*, March 10, 2005.

2. Children breathe in more air per pound of body weight than adults.

3. There has, for example, been no increase in the overall number of hurricanes and cyclones in the past hundred years.

4. One study says wind speed would rise by 5 percent for every 1 degree Celsius of warming in tropical ocean surface temperatures (tropical oceans have warmed 0.5°C in the past 50 years). Kerry Emanuel, "Increasing Destructiveness of Tropical Cyclones over the Past 30 Years," *Nature*,

August 4, 2005. Wetter, more intense hurricanes. Kevin Trenberth, "Uncertainty in Hurricanes and Global Warming," *Science*, June 17, 2005. Another study showed there has been an 80 percent increase in intense hurricanes over the past 35 years. P.J. Webster et al, "Changes in Tropical Cyclone Number, Duration, and Intensity in a Warming Environment," *Science*, September 16, 2005.

5. Bjørn Lomborg, *The Sceptical Environmentalist: Measuring the Real State of the World* (Cambridge: Cambridge University Press, 2001), 299.

6. National Oceanographic Data Center, Maryland, published in *Geophysical Research Letters*, cited in *Guardian Weekly*, February 4-10, 2005.

7. Symposium, May 2004, Paris, organized by UNESCO's Intergovernmental Oceanic Commission and International Council for Science's committee on oceanic research, http://ioc.unesco.org/iocweb/co2panel/highoceanco2.htm; report issued by Royal Society, June 30, 2005, quoted in *New York Times*, July 1, 2005.

8. Edward Goldsmith and Caspar Henderson, "Global Warming Threatens the World Economy," in Mary Williams, ed., *Is Global Warming a Threat?* (San Diego: Greenhaven, 2003), 68.

9. The speed of the Kangerdlugssuaq Glacier in eastern Greenland, one of the island's largest, increased 210 percent between 2000 and 2005, and it is now moving toward the sea at a rate of 12 kilometers a year. Eric Rignot and Pannir Kanagaratham, "Changes in the Velocity Structure of the Greenland Ice Sheet," *Science*, February 17, 2006. On the Antarctic Peninsula, 87 percent of glaciers have retreated in the past sixty years. A.J. Cook et al, "Retreating Glacier Fronts on the Antarctic Peninsula over the Past Half-Century," *Science*, April 22, 2005.

10. Bill McKibben, "Worried? Us?" This Overheating World, *Granta*, Fall 2003.

11. Jonathan Weiner, *The Next One Hundred Years: Shaping the Fate of Our Living Earth*, (New York: Bantam, 1990), 106.

12. Edward Goldsmith, "Massive Campaign Needed to Slow Down Global Warming," *CCPA Monitor*, Canadian Centre for Policy Alternatives, October 2004.

13. Jeffrey E. Lovich, "Turtles and Global Climate Change," *Impact of Climate Change and Land Use in the Southwestern US*, US Geological Survey, November 25, 2003, http://geochange.er.usgs.gov/sw/impacts/biology/turtles.

14. Government of Canada, www.climatechange.gc.ca/ Manitoba.

15. Steven Loranger, "The Water Century," *The Globalist*, March 4, 2005.

16. Peter Schwartz and Doug Randall, "An Abrupt Climate Change Scenario and Its Implications for United States National Security," October, 2003, www.environmental defense.org/documents/3566_AbruptClimateChange.pdf.

17. Jacqueline McGlade, executive director, European Environmental Agency, quoted in *Globe and Mail*, May 21, 2005.

5 The Frightening Numbers

1. Niels Bohr, physicist and Nobel Prize winner, 1885-1962.

2. Agnes Sinai, "The Cleanliness Business," *Le Monde Diplomatique*, March 2006.

3. *The Economist*, "Changing Science," December 10, 2005; John Houghton, *Global Warming: The Complete Briefing*, Third Edition (Cambridge, Cambridge University Press, 2004), 69.

4. At the current rate of increase, the International Energy Association (IEA) expects greenhouse gas emissions to

increase by 60 percent by 2030. Patrick Brethour, "IEA Says Policies Inadequate, Emissions Will Jump 60%," *Globe and Mail,* October 27, 2004.

5. The draft 2007 IPCC report estimates that if carbon dioxide levels double, average global temperatures will rise 2.0 to 4.5°C by 2100. Jim Giles, "US Posts Sensitive Climate Report for Public Comment," *Nature,* May 4, 2006; Richard A. Kerr, "How Hot Will the Greenhouse World Be?" *Science,* July 1, 2005.

6. Environment Canada, www.msc-smc.cc.gc.ca/media/top10 /2004_e.htm/#topten.

7. The IPCC range is 9 to 88 cm (4 to 35 inches). IPCC *Climate Change 2001: The Scientific Basis* (Cambridge: Cambridge University Press, 2001), 671.

8. Bjørn Lomborg, *The Sceptical Environmentalist: Measuring the Real State of the World* (Cambridge: Cambridge University Press, 2001), 289.

9. William K. Stevens, *The Change in the Weather: People, Weather, and the Science of Climate* (New York: Delta, 1999), 182.

10. Susan Joy Hassel et al, "Impacts of a Warming Arctic," Arctic Climate Impact Assessment (ACIA), www.amap.no /acia/.

11. F.S. Chapin et al, "Role of Land-Surface Changes in Arctic Summer Warming," *Science,* October 28, 2005.

12. Mark Lynas, "Hot News," This Overheating World, *Granta,* Fall 2003.

13. Quirin Schiermeier, "A Sea Change," *Nature,* January 16, 2006; Houghton, 136.

14. Although most climate models do not predict a significant slowdown until the end of the century, a recent study reports that circulation in the north Atlantic has weakened by 30 percent over the past fifty years. H.L. Bryden et al,

"Slowing of the Atlantic Meridional Overturning Circulation at 25°N," *Nature*, December 1, 2005.

15. Collapse of fisheries will be largely due to the depletion of North Atlantic plankton stocks by up to 51 percent; plankton form the base of the marine food chain. Andreas Schmittner, "Decline of the Marine Ecosystem Caused by a Reduction in the Atlantic Overturning Circulation," *Nature*, March 31, 2005.

16. Filippo Giorgi, "Climate Change Hot-Spots," *Geophysical Research Letters*, April 21, 2006.

17. Houghton, 170.

18. Mark Lynas, *High Tide: The Truth About Our Climate Crisis* (New York: Picador, 2004), 23-24.

19. Moving 500 meters up a mountain is like moving 250 kilometers north. Jonathan Weiner, *The Next One Hundred Years: Shaping the Fate of Our Living Earth* (New York: Bantam, 1990), 178.

20. United Nations Framework Convention on Climate Change (UNFCCC), *Beginner's Guide*, 1994, www.unfccc.org/resources/beginner.html.

21. Mark Maslin, *Global Warming: Causes, Effects and the Future* (Osceola, Wisconsin: Voyageur, 2002), 53.

22. Queensland Centre for Marine Studies, www.marine.uq.edu.au.

23. Number of species "committed to extinction" is estimated at 15 to 37 percent. Chris D. Thomas et al, "Extinction Risk from Climate Change," *Nature*, January 8, 2004.

24. Houghton, 187.

25. (Northwest Passage) "Impacts of a Warming Arctic," Arctic Climate Impact Assessment, 2004, www.amap.no/acia/. (Polar bears) *National Geographic*, September 2004. (Great Lakes) Government of Canada, www.adaptation.nrcan.gc.ca. (Missouri and Upper Mississippi basins) Houghton,

160. (Louisiana) *National Geographic*, October 2004. (British Columbia) Government of British Columbia, www.for.gov.bc.ca/hfp/mountain_pine_beetle/. (Peru) International Strategy for Disaster Reduction, www.eird. org. (Southern Europe) Houghton, 130. (Himalayas) Lynas, 238. (China) Worldwide Fund for Nature, quoted in Houghton, 152. (South Pacific) Michael C. Howard, ed., *Asia's Environmental Crisis* (Boulder, Colorado: Westview, 1993), 13-14. (Nile) Lynas, 113-14. (Coral reefs) Andrea G. Grottoli et al, "Heterotrophic Plasticity and Resilience in Bleached Corals," *Nature*, April 27, 2006.

26. Avoiding Dangerous Climate Change, www.stabilisation 2005.com.

27. Hadley Centre, *Uncertainty, Risk and Dangerous Climate Change: Recent Research on Climate Change Science from the Hadley Centre*, December 2004, www.metoffice.com/ research/hadleycentre.

28. "Impacts of a Warming Arctic," Arctic Climate Impact Assessment (ACIA).

29. Julian A. Dowdeswell, "The Greenland Ice Sheet and Global Sea-Level Rise," *Science*, February 17, 2006; Jonathan T. Overpeck et al, "Paleoclimatic Evidence for Future Ice-Sheet Instability and Rapid Sea-Level Rise," *Science*, March 24, 2006.

30. "More people are projected to be harmed than benefited by climate change, even for global mean increases of less than a few °C." IPCC, *Climate Change 2001: Impacts, Adaptation, and Vulnerability*, (Cambridge: Cambridge University Press, 2001), 8.

6 The Tough Questions

1. Jared Diamond, *Collapse: How Societies Choose to Fail or Succeed* (New York: Viking, 2005), 521.

2. Leopold Center for Sustainable Agriculture, www.leopold. iastate.edu; Worldwatch Institute, www.worldwatch.org.

3. Energy Information Administration, 2003 data, www.eia. doe.gov/neic/brochure/infocard01-htm.

4. Population Reference Bureau, www.prf.org.

5. International Energy Association, www.iea.org.

6. John Whitelegg, "The Global Transport Challenge," *Open Democracy*, April 26, 2005; World Book 2001; Peter Gwynne, "Green Cars Move into Top Gear," *Physics World*, July 2002.

7. Mark Maslin, *Global Warming: Causes, Effects and the Future* (Osceola, Wisconsin: Voyageur, 2002), 22.

8. Tim Grant and Gail Littlejohn, *Teaching About Climate Change: Cool Schools Tackle Global Warming* (Gabriola Island, BC: New Society, 2001), 52.

9. *Globe and Mail*, "To Lose Pounds, Even Wires Get Skinny," June 2, 2005.

10. Linda McQuaig, *It's the Crude, Dude: War, Big Oil, and the Fight for the Planet* (Toronto: Random House, 2004), 5.

11. Clive Ponting, *A Green History of the World: The Environment and the Collapse of Great Civilizations* (New York: Penguin, 1991), 213.

12. United Nations Framework Convention on Climate Change *Beginner's Guide*, 1994, www.unfccc.org/resource /beginner.html.

13. World Bank, www.worldbank.org/data/wdi2004/pdfs/ table205.pdf.

14. Fareed Zakaria, *The Future of Freedom* (New York: Norton, 2003), 82; Matt Walker, "A Nation Struggling to Catch Its Breath," *New Scientist*, April 29, 2006.

15. Geoffrey York, "China's Unquenchable Thirst," *Globe and Mail*, May 21, 2005.

16. *The Economist*, "Power to the People," February 11, 2006.

17. *National Geographic*, May 2004.
18. Tom Burke, "Don't Push Bush on Kyoto," *Guardian Weekly*, February 4-10, 2005.
19. Andrew Heintzman and Evan Solomon, eds., *Fueling the Future: How the Battle over Energy Is Changing Everything* (Toronto: Anansi, 2003), 279.
20. Heintzman, 135.
21. Michael Williams, *Deforesting the Earth* (Chicago: University of Chicago Press, 2003), 496.
22. Britaldo Silveira Soares-Filho et al, "Modelling Conservation in the Amazon Basin," *Nature*, March 23, 2006.
23. *Chicago Tribune*, March 12, 1989, quoted in Williams, 499.

7 Facing the Music

1. Worldwatch Institute, "Climate Change: Reducing the Threat of Climate Change in the US: A Survey of Activities," 2006, www.worldwatch.org/features/climate/activities/.
2. Margaret Beckett and Patricia Hewitt, British Secretaries of State for Environment, Food and Rural Affairs and for Trade and Industry, "Naysayers, Take Note: Britain Is Pro-Kyoto and Prospering," *Globe and Mail*, March 15, 2005.
3. Committee on Science, Engineering and Public Policy (COSEPUP), *Policy Implications of Greenhouse Warming: Mitigation, Adaptation, and the Science Base* (Washington, DC: National Academy of Sciences, 1992), 201.
4. The Climate Group, *°C*, June 18, 2006, www.theclimate group.org/index/php?pid=400.
5. "The Greening of General Electric," *The Economist*, December 10, 2005.
6. James Gustave Speth, "The Single Greatest Threat,"

Harvard International Review, Summer 2005.

7. Oregon Department of Energy, "Oregon Carbon Dioxide Emission Standards for New Energy Facilities," www.oregon.gov/ENERGY/SITING/docs/ccnest.pdf.

8. Charles Petit, "Power Struggle," *Nature*, November 17, 2005.

9. Andrew Heintzman and Evan Solomon, eds., *Fueling the Future: How the Battle over Energy Is Changing Everything* (Toronto: Anansi, 2003), 101.

10. Scottish Executive Renewable Policy Document, "Securing a Renewable Future: Scotland's Renewable Energy," www.scotland.gov.uk/publications/2003/03/16850/20554.

11. Janet L. Sawin, "Run with the Wind," *New Internationalist*, January 2004; European Wind Energy Association, www.ewea.org/doc/20gw%20briefing.pdf.

12. Canadian Wind Energy Association, www.canwea.ca.

13. Arno Kopecky, "Water to Burn," *The Walrus*, December/January 2006.

14. Geothermal Education Office, http://geothermal.marin.org/map/usa.html.

15. Shell press release, October 13, 2003, www.shell.com.

16. *Ecologist*, July/August 2005, 11.

17. Matt Walker, "A Nation Struggling to Catch Its Breath," *New Scientist*, April 29, 2006.

18. Jared Diamond, *Collapse: How Societies Choose to Fail or Succeed* (New York: Viking, 2005), 294.

19. Heintzman, 310.

20. The campaign is called Cool Biz and officials say the country could save 1.9 million barrels of oil a year if office thermostats were set at 28°C instead of 25°C, as well as helping Japan to meet its Kyoto requirements (6 percent below 1990 levels by 2012). *Globe and Mail*, "Japan Saves Energy, Loosens Up with 'Cool Biz' Campaign," June 1, 2005.

21. John Houghton, *Global Warming: The Complete Briefing*, Third Edition (Cambridge: Cambridge University Press, 2004), 279.

22. Tim Grant and Gail Littlejohn, *Teaching About Climate Change: Cool Schools Tackle Global Warming* (Gabriola Island, BC: New Society, 2001), 52.

23. Richard Adams, quoted in Heather Mallick, "When Aviation Falls to Earth," *Globe and Mail*, January 29, 2005.

24. Howard Geller, "Compact Fluorescent Lighting," American Council for an Energy Efficient Economy, www.aceee.org.

25. "Energy Matters," *Canadian Geographic*, May/June 2003.

26. The city has a 200-kilometer network of bicycle and pedestrian paths. www.velomondial.net/velomondial/2000/pdf/pacheco.pdf.

27. *The Economist*, "A New Train Set," March 25, 2006.

28. Houghton, 299.

29. (Iceland) A. Kopecky, *The Walrus*, December/January 2006. (Philadelphia) Rohm and Haas, www.rohmhaas.com/roof/coolroof. (California) "California Hydrogen Highway Funding," May 10, 2005, www.hydrogenhighway.ca.gov. (Los Angeles) Charles Petit, "Power Struggle," *Nature*, November 17, 2005. (Chile) Dave Ebner, "Trans-Alta, Chilean Firm Strike Emissions Swap Deal," *Globe and Mail*, August 25, 2004. (Denmark) Greenpeace newsletter, Summer 2004; www.windenergy.org. (Europe) *National Geographic*, August 2005. (Mongolia) Houghton, 299. (Yangtze River) Emma Young, "Flood Warnings for the Yangtze Will Save Lives," *New Scientist*, September 17, 2005. (Taipei) World Car Free Day, www.worldcarfree.net/wcfd/; Mobility Week Europe, www.mobilityweek.europe.org/part/campaign_presentation_2004.html.

30. Diamond, 497, 515-16.

31. The Kyoto Protocol had its origins at the UN Conference

on Environment and Development, held in Rio de Janeiro, Brazil, in 1992. Known as the Earth Summit, the conference was attended by 30,000 participants from 178 governments. The Earth Summit identified climate change as one of the three major issues threatening the environ ment (the others were biodiversity and forests). Participants drafted the UN Framework Convention on Climate Change, a voluntary agreement, now ratified by 188 countries, to lower the emission of greenhouse gases, especially carbon dioxide. But five years later, greenhouse gas emissions had actually risen overall.

It was clear that the international community needed a formal, legally binding agreement that would include specific targets for reducing greenhouse gases. In December 1997, this agreement was drafted in Kyoto, Japan, at a conference attended by 6,000 delegates from more than 160 countries. The Protocol would become legally binding after at least 55 countries, responsible for at least 55 percent of the greenhouse gas emissions produced in 1990, agreed to reduce their emissions. Because the US, which is responsible for more than one-third of the world's green house gas emissions, refused to sign, almost all the remaining industrialized countries of the world had to come on board in order for the treaty to become law. This finally happened in November 2004, when Russia ratified Kyoto.

32. International Energy Agency, cited in *The Economist*, "Don't Despair," December 10, 2005.

Glossary

albedo The brightness of a surface that determines how much light the surface reflects rather than absorbing it as heat. Snow has a high albedo; dark rock and vegetation-covered surfaces have a low albedo because they absorb sunlight and turn the energy into heat.

atmosphere The gases surrounding the earth.

carbon cycle The circulation of carbon in its various forms between the atmosphere, the land and the ocean.

carbon sink Something (oceans, plants, etc.) that removes carbon from the atmosphere and stores it.

chlorofluorocarbons (CFCs) Human-made compounds used in refrigeration, foaming agents and aerosol sprays. CFCs destroy atmospheric ozone. They are also powerful greenhouse gases.

climate The average weather that prevails in a certain area over a long period of time.

cyclone A low-pressure system with intense rain and wind activity occurring over tropical or sub-tropical waters. In the Atlantic Ocean and in the Pacific Ocean east of the International date line such storms are called hurricanes; in the Pacific Ocean west of the date line they are called typhoons; in the Indian Ocean they are called cyclones.

El Niño A natural phenomenon that occurs when a large stretch of the southern Pacific Ocean becomes superheated and flows toward South America, bringing intense storm activity and disrupting weather patterns around the world.

fossil fuel Coal, oil or natural gas formed from the decayed remains of ancient animals and plants.

fuel cell A device that produces electricity from a chemical reaction of oxygen and hydrogen.

General Circulation Models (GCMs) Computer programs designed to predict what will happen to the earth's climate.

geothermal energy Energy obtained from heat deep below the earth's crust.

global warming The gradual increase in the overall average temperature of the earth's surface, caused by the greenhouse effect.

Greenhouse Effect Warming of the earth's atmosphere caused by gases that trap heat emitted from the ground.

greenhouse gas One of several gases, including water vapor, carbon dioxide and methane, that trap heat, creating the greenhouse effect.

hydrologic cycle The circulation of water in its various forms between the atmosphere, the land and the oceans.

hydro power The use of water power to generate electricity.

ice age A period in the earth's history when temperatures were low and much of the earth's surface was covered by snow and ice. Ice ages are interrupted by warmer interglacial periods.

Industrial Revolution The change from a primarily farming to industrial society that occurred in Europe and North America during the nineteenth century.

Intergovernmental Panel on Climate Change (IPCC) The international scientific body that assesses global warming.

jet stream A band of westerly winds that prevails in the mid latitudes of each hemisphere.

monsoon Seasonal periods of heavy rainfall in sub-tropical regions.

permafrost Land that is frozen year round.

renewable energy An energy source (such as sun or wind) that cannot be used up.

stratosphere The layer of atmosphere that sits between 10 to 50 kilometers (6 to 31 miles) above the earth's surface.

thermohaline circulation (THC) Circulation in the world's oceans, driven by different water densities determined by temperature and salinity.

troposphere The layer of lower atmosphere up to a height of 10 kilometers (6 miles).

tundra Treeless habitat of the Arctic.

weather The day-to-day conditions in the atmosphere in a specific place.

For Further Information

Dozens of articles, books and websites were used in the research of this book. Many are listed in the notes on pages 115 to 129. But the science of global warming is changing so rapidly that sources are often quickly outdated. Reliable websites are often the best source for current information (look for the dates the information was posted; they can be hard to find).

Here are a few:

Intergovernmental Panel on Climate Change (IPCC)

Established by the World Meteorological Organization and the United Nations Environment Programme to assess the scientific, social and economic issues surrounding human-induced climate change.

www.ipcc.ch

United Nations Environment Programme (UNEP)

Outlines strategies for addressing climate change, based on the UN Framework Convention on Climate Change (UNFCCC) and the Kyoto Protocol.

www.unep.org/themes/climatechange

Science and Development Network

Covers scientific and technological issues from the perspective of the developing world. Pages devoted to climate change and environmental issues include news, recent studies and editorials.

www.scidev.net/index.cfm

Climate Action Network

More than 340 NGOs working to promote government and individual action to limit human-induced climate change to ecologically sustainable levels; provides links to offices around the world for information on regional concerns and initiatives.
www.climatenetwork.org

Climate Ark and Climate Wire

Two search engines and information services that direct users to climate change websites and recent news releases.
www.climateark.org
www.climatewire.org

Acknowledgments

The author and publisher would like to thank Maria Luisa Crawford, Greg Flato, Danny Harvey and John Houghton for their helpful comments and critique of the manuscript. Thanks also to Jessica Abraham, Keith Abraham, Ryan Bergen and Leon Grek for editorial assistance, fact-checking and additional research.

Index